Introduction to
TOXICOLOGY
and FOOD

Introduction to
TOXICOLOGY
and FOOD

Tomris Altuğ

CRC PRESS

Boca Raton   London   New York   Washington, D.C.

Library of Congress Cataloging-in-Publication Data

Altug, Tomris
    Introduction to toxicology and food : toxin science, food toxicants, chemoprevention /
        p. cm.
    Includes bibliographical references and index.
    ISBN 0-8493-1456-9
    1.Food--toxicology. 2. Toxicology. 3. Functional foods. I.Title.
    RA 1258 .A425 2002
615.9′54—dc21   2002067066
                                                             CIP

Visit the CRC Press Web site at www.crcpress.com

© 2003 by CRC Press LLC

No claim to original U.S. Government works
International Standard Book Number 0-8493-1456-9
Library of Congress Card Number 2002067066
Printed in the United States of America   2 3 4 5 6 7 8 9 0
Printed on acid-free paper

DEDICATION

To my dear mother and brother,
who were always there when I needed them

PREFACE

Food science is a multidisciplinary subject with relationships to such sciences as chemistry, biochemistry, microbiology, nutrition, engineering and toxicology. The safety of foods is one of the most important quality characteristics that affect their acceptability by consumers, as well as the legal authorities. The different varieties of foods available to today's consumers may contain several chemicals that could affect their safety.

Food is a complex structure that consists of natural chemical components such as proteins, carbohydrates, fats, minerals, vitamins, enzymes, etc. Most of these substances that compose the chemical structure of foods are necessary for the human being to survive and do not cause toxic symptoms at normal levels of consumption. However, there are also some substances, such as cyanides, oxalates, saponins, etc., in foods that, while naturally present, may have toxic effects when consumed at normal levels of use.

Besides these natural components, several chemicals may be incorporated, either intentionally or not, during the production, processing, handling, packaging, storage or transportation of foods. The intentional addition of chemicals for specific purposes in foods goes back to early history with the application of practices like salting, smoking and pickling for preservation of meat and fish, improving flavor by using spices and herbs and attracting consumers by the addition of colors. The use of chemical additives in foods for various purposes such as preservation, increasing shelf life, improving sensory properties, producing foods for special groups of consumers and aiding in the different stages of food manufacture, is gaining wide acceptance all over the world.

On the other hand, foods may be contaminated by unintended chemicals during various stages of production, beginning from the cultivation of the crops or the breeding of animals, throughout processing operations up to the finished products and even during the transportation, storage and sales stages of such foods in the market. The toxicological problems related to these foreign chemicals come mainly from contaminants whose presence cannot be avoided in many cases. The health hazards that may result by their consumption can only be inhibited by effective food control techniques.

Intentional chemical additives are authorized only if no harmful effects of any kind can be shown after extensive toxicological testing, and many food additives have undergone more rigorous testing than some foods or food components. A good deal of work is performed in establishing specifications and limits of use of these chemicals in food, and the benefits of intentional additives are gaining importance with food technologists every day. Therefore, advances in food technology have created the need for gaining information about the science of toxicology and toxicants in foods.

Today, many consumer groups encourage the production of foods that are safe from a toxicological point of view. Consumers are also interested in foods that can be used for functional purposes and that supply the necessary nutrients to survive and prevent some diseases. Among the foods or food components that are consumed for the latter purpose, the substances that may have antimutagenic or anticarcinogenic effects are gaining importance every day.

Food toxicology is a science that deals with the nature, sources and formation of toxic substances in foods, their harmful effects, the mechanism of toxic effects and the identification of the safety of these substances. A primary objective of food toxicology is an understanding of the nature and properties of all toxic substances in foods as well as the nature and magnitude of the hazard they present under different conditions. Also important is determining their safety limits. The first part of this book covers the general concepts and principles in toxicology by describing its history and branches, toxic doses, stages of toxication, effect mechanism of toxins and toxicity tests. In the second part, the substances in our foods that have toxicological significance, such as natural sources of toxicants, contaminants and food additives, are described. The third part involves information about the food components and foods themselves, which can be expressed as "chemopreventors" due to their antimutagenic or anticarcinogenic effects *in vitro* and *in vivo* systems, as well as in epidemiological studies.

I sincerely hope that this book will make a contribution to the wide area of the multidisciplinary science food toxicology.

Tomris Altuğ

ACKNOWLEDGMENTS

I would like to thank my colleague Professor Dr. Tavman and my dear colleagues in the Food Engineering Department of Ege University, who encouraged me in the preparation of this book. I would also like to express my gratitude to Dr. Eleanor Riemer and to CRC Press for their enthusiasm and help in the publication of this book.

AUTHOR

Tomris Altuğ is a professor of food science in the Food Engineering Department of Ege University, Turkey. After completing her secondary and high school education in American Girls' School in Izmir, she received B.S. and M.S. degrees in Chemistry from Ege University. In 1979, she began working as a research assistant in the Food Engineering Department of Ege University and she received her Ph.D. in Food Science. Dr. Altuğ has taught several courses in food science including Food Toxicology and Food Quality Control and supervised the master's and doctorate thesis of several students in the Food Engineering Department of Ege University. In 1984, on a British Council scholarship, she attended a training program on aflatoxins and mycotoxins in the Tropical Development and Research Institute in London. In 1989, on a USDA scholarship, she visited the University of Wisconsin-Madison to join a research program on degradation of aflatoxin B_1 in dried figs and, during her stay, also visited Texas A&M and California-Davis Universities. Dr. Altuğ has published several articles and books related to food science. She has also received Tubitak Scientific Motivation awards for some of her studies. Dr. Altuğ's major areas of interest are related to different aspects of food quality, extending from food toxicology to the sensory properties of foods and focusing on the philosophy of "production of safe foods that are palatable to the consumer."

CONTENTS

1

GENERAL INFORMATION ON TOXICOLOGY

INTRODUCTION

In a general sense, toxicology can be defined as "toxin science" or the "science of poisons." The main divisions of the study of toxicology involve the sources of toxins, physical, chemical and biological properties of toxins, toxic doses, the changes that occur in living organisms and their effects, treatment of toxic diseases, isolation of toxins, analysis of toxins and regulations about toxins. Toxicology frequently deals with foreign substances that are apart from those necessary for the normal metabolism of a living organism. In recent years, these foreign substances have been termed "xenobiotics."

Toxicology also deals with the substances that are necessary for the body, such as some hormones, amino acids, vitamins and exogenous substances, such as food additives, that might have toxic effects at high levels. Many toxic events have occurred in recent years due to the increased use of industrial, agricultural and household chemical substances, and with the use of nuclear energy. These harmful effects concern the whole biosphere as well as human beings. The investigation of these chemical substances in biological systems and in the environment, and the methodology related to this investigation are the subjects of the science of chemistry, while their metabolism, effects and changes at molecular level are the subjects of biochemistry. The investigations of the toxic effects that occur as the results of the use of these chemicals, as well as the treatment and research on the assurance of their use, are related to the sciences of medicine, agriculture and food. Toxicology is a multidisciplinary science with close relationships with other sciences such as pharmacology, immunology, biology, pathology, physiology, chemistry, biochemistry, food and public health.

THE HISTORY OF TOXICOLOGY

Toxin science is as old as the primitive societies. Toxins can be defined as substances that may cause adverse health effects under the conditions in which they are produced or used. Archeological research has shown that ancient people knew about toxins of plant, animal and mineral origin. *Native Americans* used the extracts of toxic plant seeds that involved toxic glycosides as weapons. At that stage of history, toxins were regarded as tools of war to be used by people to defend themselves against their enemies. The papyrus scrolls of Egypt are known to be the oldest sources of information about toxicology. Ebers papyrus, written in 1552 B.C., the oldest written medical record, involves more than 800 formulas and explains the production of toxins such as belladonna and opium alkaloids, lead, antimony and copper. Rational medicine began in the time of the ancient Greeks. Hippocrates (460–315 B.C.) used toxins in the treatment of certain diseases while adding new toxic substances in the area of medicine. He also established the basis of industrial hygiene and toxicology. The early Romans used the toxins for political purposes, when emperors poisoned and killed their enemies by using toxic substances. During the Middle Ages and prior to the Renaissance period, poisoning became something of an art; such events were widespread in France, Italy, Holland and England. Criminal poisoning continued during the 18th and 19th centuries. In our time, although toxins and poisoning events still have importance, the point of view differs from that of criminal poisoning.

A very important scientist in the late Middle Ages, Paracelsus, was regarded as a "Renaissance man" in the history of science and medicine. Paracelsus was the creator of the basic scientific discipline of toxicology, to the development of which he made valuable contributions. He created the focus on the "toxicon," the toxic agent, as a chemical entity, and emphasized the study of the relation of chemical structure to toxicity. He declared that experimentation is essential in examination of the responses of living organisms to chemicals and that the distinction between therapeutic and toxic properties of chemicals should be well identified. He also stated that the degree of specificity of chemicals and their therapeutic or toxic effects could be confirmed by regarding the doses of these chemicals. This was the first significant description of the dose–response relation, which is regarded as the building block of toxicology science. The studies of Paracelsus were extended from detection of accidental or intentional poisonings through a series of environmental factors and occupational diseases of workers. His interests and activities were mainly related to forensic, industrial and environmental toxicology.

MODERN TOXICOLOGY

Matthieu Joseph Bonaventura Orfila (1787–1853), a Spanish scientist, is known to be the establisher of modern toxicology. He was a medical doctor who specialized in chemistry and physiology after completing his education in medicine. He was the first person to find a systematic relationship between the chemical and biological effects of toxins. He performed several experiments on animals (dogs) investigating the effects of poisons. He distinguished toxicology as a distinct discipline by defining it as the "study of poisons." His main contribution to toxicology was discovering the fact that toxins not only accumulate in the stomach, but are distributed to several organs of the body after being absorbed by the gastrointestinal system. Until this time, the toxins were searched for only in the stomachs of the victims. Orfila introduced chemical analysis as legal evidence in a poisoning event that resulted in death, as well as improving many methods for identification of toxins. In this way, Orfila established the basis of analytical toxicology and forensic toxicology, which are the main divisions of modern toxicology.

In the 20th century, developments occurred rapidly, and the effects of several toxic substances, including their mechanisms, were investigated. The concept of an antidote for specific toxic substances and the treatments for toxic diseases were evolved. Developments that led to the discovery and understanding of toxic substances for use by man included the discovery and study of DDT and organophosphate insecticides. The rapid advancement in toxicology that occurred through the studies related to analytical sensitivity has created the necessity of regarding toxicology as a science and of subdividing it into several branches.

BRANCHES OF TOXICOLOGY

Analytical Toxicology

This branch, also called chemical toxicology, deals with the isolations of toxins from the biological material, definitions of toxins, and investigations of qualitative and quantitative methods for the analyses of toxic substances. Generally, toxic substances are found in trace amounts (ppm, ppb) in the biological substrates, so the micromethods used for their analysis must be sensitive, accurate and precise. The analysis in chemical laboratories of food contaminants, food additives and some natural toxicants of foods are related to this branch of toxicology.

Biochemical Toxicology

This branch, which examines the reaction of toxic substances in the living organism at a molecular level, forms the basis of all the other branches of toxicology. The explanation of the effect mechanism of toxic events on organisms is necessary in the production of economic toxins, treatment of toxications and analysis of toxic substances. Biochemical toxicology investigates the changes in xenobiotics in the living organism (absorption, distribution, metabolism and excretion) and examines these changes at a molecular level. The biochemical changes that occur in living organisms following consumption of contaminants, food additives and natural food toxicants are observed before establishing their maximum levels in foods. The responses to the effects of chemical substances that occur in living organisms are called "behavior toxicology" and are regarded as another branch of toxicology.

Economic Toxicology

This branch of toxicology deals with the selective effects of chemical substances on biological tissues. Using the knowledge of these effects, several drugs, food additives and pesticides have been developed. Many natural and synthetic chemical substances show a toxic effect on microorganisms like bacteria, virus and fungi, and retard or inhibit their growth while not affecting the human organism. This is known as "selective chemical effect" or "selective toxicity" and it can be defined as being toxic or showing toxic effect of a specific kind to a tissue, organ or cell. This effect is used in the production of economic toxins known as pesticides, drugs and food additives. In a commercial application, pesticides eliminate the weeds, microorganisms, insects and other pests that damage plants. The best examples of economic toxins are food additives that are used as antimicrobials and antioxidants for preventing foods from the harmful effects of microorganisms and oxygen, while doing no harm to humans at their permitted levels of use.

Environmental Toxicology

This branch of toxicology deals with the toxicological events that occur as a result of the exposure of human beings to the chemical substances that pollute the environment. The chemical substances that are used for various purposes pollute air, water, soil and our foods during their production stages or by way of industrial effluents. This pollution may have harmful effects on all living organisms, including human beings. The pesticides and fertilizers used in agriculture, trace metals, asbestos, polychlorinated biphenyls and polybrominated biphenyls (PCBs and PBBs) and many other chemical substances used in industry pollute the water and soil. The burned products evolved from industrial smokestacks and house chimneys, exhaust gases from motor vehicles, polycyclic aromatic hydrocarbons (PAHs) resulting from burning of coal and forest fires, pollute the air. Environmental toxicologists investigate the toxicity and toxic effects of these pollutants and attempt to prevent the harmful effects of these substances on public health by establishing regulations on their maximum permitted levels (MLs) in water, soil and air. Environmental toxicology is closely related with food toxicology because our foods may also be contaminated by pollutants such as PCBs, PAHs, asbestos, mercury, cadmium and several other chemicals. Extensive studies are performed to prevent the contamination of our foods by these substances, and laws in different countries of the world regulate the presence of these toxicants in foods.

Forensic and Clinical Toxicology

This branch of toxicology, which began with Orfila, deals with the harmful effects of chemical substances on humans and animals and examines poisonings from the legal point of view. In the evaluation of such poisonings, the determination of the cause–effect relation of the chemical substance being exposed is important. This evaluation is performed by analyzing the levels of the toxic substance in biological materials such as blood, urine, tissue or organs and by determining the dose–response relations. Thus, forensic toxicology is closely related to analytical toxicology. Forensic toxicology deals with both accidental and intentional poisonings, while clinical toxicology works apart from forensic toxicology in dealing with the treatment of toxic diseases. It is related to food toxicology in cases of food poisoning events resulting by ingestion of food-borne toxicants. Clinical toxicology deals with the diseases caused by, or uniquely associated with, toxic substances. The clinical toxicologist's main responsibility is to reduce the injury of toxic substances to the victims by treating toxic diseases, thus, this branch of toxicology is mainly related to the area of medicine.

Genetic, Reproductive and Developmental Toxicology

Genetic toxicology deals with the interaction of chemical and physical agents with the process of heredity, including studies of mutagenesis and carcinogenesis. Reproductive toxicology, which involves studies of teratogenesis, is concerned with possible effects of substances on the reproductive process. Developmental toxicology can be categorized as a subbranch of reproductive toxicology. Today, much new information is available about the origins of developmental disorders resulting from chemical exposure. This branch examines the nature and application of chemicals, including environmental pollutants that might induce developmental toxicity and the route by which they can reach the unborn. It also concerns the concentration of the chemicals and their metabolites in the embryo or fetus.

Juridical Toxicology

The increase of toxicological events that occur because of the excessive amounts of chemical substances produced and their wide use, led to the necessity of defending the public and the environment from the producers of such substances. Juridical toxicology deals with the legal procedures and regulations for prevention of the harmful effects of chemical substances on living organisms, both human and environmental. This subject interests producers and consumers alike from economic and health points of view, while it is related to the governments and local administrations from a legal point of view. The establishment of regulations on food contaminants, food additives and natural food toxicants can be given as examples of the activities related to juridical toxicology.

Industrial Toxicology

This branch of toxicology deals with the injurious potentials and the dose–response relationships of the substances encountered by workers: raw materials, intermediates and finished products. In our time, the consumption of a wide variety of chemical substances increases with the growing human population. The people who work in the production of drugs, pharmaceutical preparations, fertilizers, pesticides, food additives, food packages and containers, plastic materials, organic substances used as war devices, organic metal compounds and nuclear chemicals are exposed daily to these toxic substances during their synthesis. The purpose of industrial toxicology is to protect the health of these people and create health assurance during their work. The acute toxicity, chronic toxicity and the special toxic effects of these chemical substances used in industry are evaluated and classified according to their toxicity levels. As the result of these studies, industrial hygiene standards are established and applied in production systems including the food industry. The industrial toxicologist must also be concerned about air, water and soil pollutants that are the results of industrial processes, thus, this branch of toxicology is closely related to environmental toxicology.

2

DEFINITION OF TOXICITY AND CLASSIFICATION OF TOXINS

TOXICITY

Toxicity can be defined as "the capacity of a substance to cause adverse health effects (injury, hazard) on a living organism." The type of toxicity that occurs can be local effects such as skin irritation, or general effects such as in impaired coordination, behavioral changes, organ structure changes or death. The toxicity of a chemical substance is related to the amount, or dose, taken into the organism. Paracelsus stated this evidence in the 15th century using the following phrase: "All substances are poisons; there is none which is not a poison. The right dose differentiates a poison and a remedy." Thus, the amount of a chemical a person is exposed to is important in determining the extent of toxicity that will occur. In toxicity tests, a wide variety of the doses of the same chemical are administered and the effects are examined by plotting the results on dose–response curves. The response of living organisms to the same chemical differs according to factors such as strains, species, age, sex and nutritional status. This can be observed from Figure 2.1, which shows the dose–response relations of two different species for the same chemical.

For practical reasons and industrial purposes, the toxicities of chemical substances are expressed by using their LD_{50} values. LD_{50} values are also expressed as "median lethal dose," and can be defined as the amount of toxin necessary to kill 50% of experimental animals. LD_{50} value indicates the degree of toxicity of various substances. Toxic doses of chemical substances are different and the lethal doses have a very wide range.

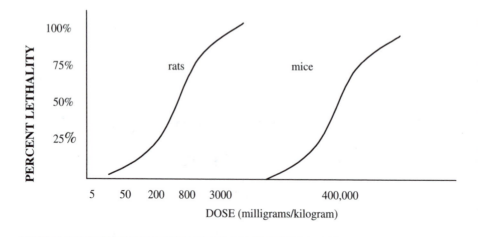

Figure 2.1 Dose–response relationships of rats and mice for the same chemical. From Kamrin, 1988, with permission of CRC Press.

Table 2.1 Toxicity Rating Chart

Toxicity Rating	Commonly Used Term	LD_{50} Single Oral Dose Rats	Probable Lethal Dose for Man
1	Extremely toxic	<1 mg/kg	A taste, 1 grain
2	Highly toxic	1–50 mg/kg	1 teaspoon, 4cc
3	Moderately toxic	50–500 mg/kg	1 ounce, 30 g
4	Slightly toxic	0.5–5g/kg	1 cup, 250 g
5	Practically nontoxic	5–15 g/kg	1 quart, 1000 g
6	Relatively harmless	>15 g/kg	>1 quart

Adapted from Derelanko, 1995.

Table 2.1 shows the toxicity rating of substances according to their approximate LD_{50} values for rats and for average adults.

Table 2.2 shows the acute LD_{50} values of several chemicals. As can be seen from the table, botulinus toxin, which is formed in foods by the anaerobic bacteria *Clostridium botulinum* is the most toxic compound among the other toxicants.

Table 2.2   Approximate Acute LD$_{50}$s of Selected Chemical Agents Found in Foods

Chemical Agent	LD$_{50}$ (mg/kg)[a]
Polybrominated biphenyls (PBBs)	21500
Ethyl alcohol	14000
Sodium chloride	4000
Ferrous sulfate	1500
Malathion	1200
Lindane	1000
2,4-D	375
Ammonia	350
DDT	100
Heptachlor	90
Arsenic	48
Dieldrin	40
Dioxin (TCDD)	0.001
Botulinus toxin	0.00001

[a] These units allow comparison with values commonly found in rat experiments.

Adapted from Klaassen and Doull, 1980 and Kamrin, 1988.

CLASSIFICATION OF TOXINS

Toxins can be defined as substances that cause adverse health effects (injury or hazard) to a living organism. Toxic substances can be classified in several ways, according to the areas of interest of the individuals who deal with the subject of toxicology.

Classification According to Their Toxic Doses

In this method of classification, the LD$_{50}$ values of the toxins are determined and their toxicities are evaluated according to the toxicity-rating chart illustrated in Table 2.1.

Classification According to the Methods of Isolation from Natural Sources

Historically, toxins were classified according to their way of isolation from natural sources of plant, animal or mineral origin. The animal sources used in these classifications included the toxins produced in the specialized organs of snakes, spiders and marine animals. Current classifications based on this approach include marine organisms because fish poisons such as ciguatoxin, saxitoxin and tetrodotoxin are caused by marine organisms in the diet of fish, and these toxins may be concentrated in the process of preparing food or protein sources. Examples of plant sources from food toxins are caffeine, yellow rice, gossypol and certain fungi, while trace metals and antibiotics can be given as examples of toxins isolated from mineral sources.

Classification According to Physical States

Toxic substances can be classified according to their physical states. Examples of this method of classification are toxins in a gaseous state, e.g., hydrogen sulfide and sulfur dioxide; toxins as vapor, e.g., benzene and hexane; toxins in aerosol form, e.g., insecticides and herbicides; and toxins in dust form, e.g., aflatoxins and asbestos powder.

Classification According to Their Use, Labels and Chemical Structures

Toxins can also be classified according to their use and labels, such as explosives, pesticides, solvents, food additives, plasticizers, etc. They can be classified according to their chemical structures, such as polychlorinated biphenyls (PCBs), polycyclic aromatic hydrocarbons (PAHs), organometallic compounds, ametallic compounds, etc. The chemical structure and the biological activity of toxins are related to each other because specific functional groups can show specific toxic effects. In addition, the isomerism in the chemical structure (optical activity and structural isomerism) can affect the biological activity of toxins.

Classification According to Physiological Effects

Toxins can be classified according to their physiological effects. In this method of classification, the tissue or target organs affected by the toxin (liver toxins, bone marrow toxins, kidney toxins), the physiological changes that occurred (central nervous system depressors, teratogenesis, carcinogenesis, mutagenesis) and biochemical effect mechanisms (toxins producing methemoglobinemia) are taken into consideration.

Classification According to Their Isolations: Qualitative and Quantitative Methods of Analysis

From the point of view of analytical toxicology, toxins are classified according to their isolations, using qualitative and quantitative methods of analysis. Toxins classified according to their isolations from biological materials are named volatile toxins, nonvolatile toxins, metallic toxins and toxic anions. They can also be classified according to their behavior of dissolving in several solvents, such as polar or nonpolar solvents. This type of classification can be modified according to the improvements and modifications in analytical chemistry methods.

3

TOXICATION

A toxic substance shows its harmful effect by entering a reaction with the living organism; this effect is called *toxication*. As a result of toxication, some symptoms appear in the biological system such as convulsion, tremor, nausea and aches in several parts of the body. Also, some breakdowns at the macromolecular level occur, which can lead to liver degenerations, leukemia, methemoglobinemia, etc. In toxication, a toxic substance passes through the following phases in showing its toxic effects (Figure 3.1).

EXPOSURE, ENTRANCE AND ABSORPTION

Exposure

The toxic effects of chemical substances can be acute or chronic depending on the period and frequency of exposure to the toxins. *Acute toxicity* occurs through being exposed to the toxic dose of a chemical substance for a short period of time (24 hours or less), but frequently. The symptoms of acute toxicity are different from the symptoms of chronic toxicity, which occur as a result of being exposed to small doses of a toxin for a long period. For example, while the main symptom of acute benzene toxication is depression of the central nervous system, in chronic benzene toxication, hematological defects occur. Toxication symptoms occur immediately in acute toxicity and the length of time before the toxified person either dies or can be saved from death is very short. Sometimes "acute" effects of a toxic dose of a substance that entered an organism in a very short period of time can be seen after a long time interval. For example, symptoms like hair loss and carcinogenic effects can occur long after being exposed to acute radiation. This late-occurring effect is defined as *delayed acute effect* or *delayed acute toxicity*.

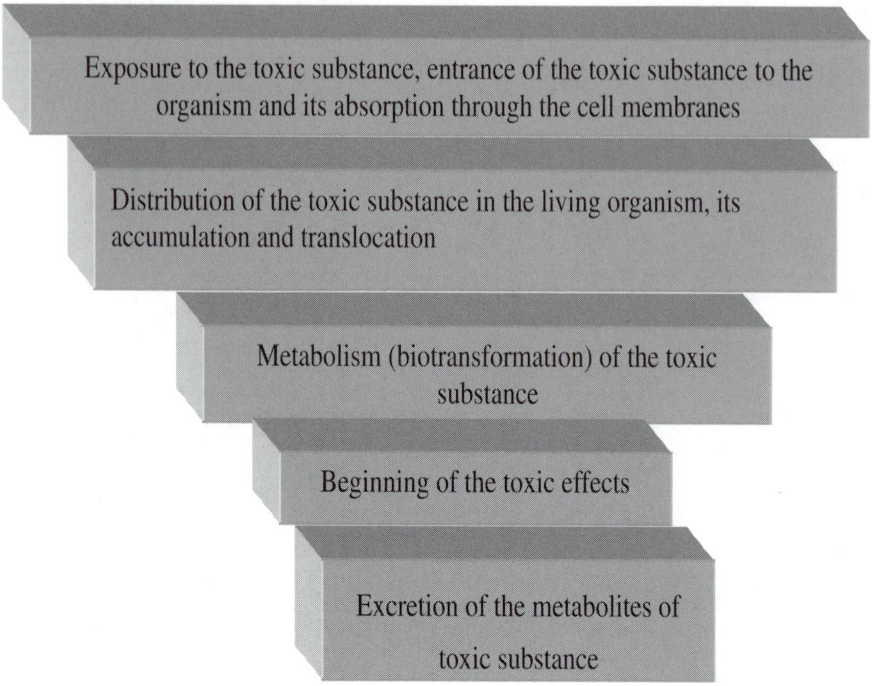

Exposure to the toxic substance, entrance of the toxic substance to the organism and its absorption through the cell membranes

Distribution of the toxic substance in the living organism, its accumulation and translocation

Metabolism (biotransformation) of the toxic substance

Beginning of the toxic effects

Excretion of the metabolites of toxic substance

Figure 3.1   Stages of toxication.

Chronic toxicity occurs as a result of being exposed to the effects of cumulative toxins. Generally, if the rate of discharge of a toxin from the organism is slower than its absorption rate, this substance shows an accumulative property. Knowledge of chronic toxicity is important for workers who are exposed to chemical substances in industry; there are various career diseases that occur as a result of chronic toxication. In addition, exposure to environmental pollutants such as DDT, PCBs, cadmium and mercury create chronic toxication, which is important from the point of view of public health. Although the symptoms of chronic exposure occur only after a long period of time, acute symptoms may also occur after each exposure to the toxic material. *Subacute toxication* may occur by subacute exposure to toxins, that is, being exposed for short periods frequently (over a period of 28–90 days). These types of toxication are met in persons who work in agriculture and use pesticides (insecticides). The symptoms of subacute toxication are identical to the symptoms of acute toxication.

Entrance

Toxins enter living organisms by way of the skin, lung, gastrointestinal system and injections. If they are to show their toxic effects, they should pass into the circulation system through the biological membrane and travel to their target points. Therefore, the toxicity of the same substance can differ at different times or in different victims, according to its means of entrance to the organism. "Parenteral" is the means of entrance of toxins by injection, while the inhalation method of entrance of toxins is important in industrial toxicity. Entrance through the skin (percutaneous) includes toxication by the use of detergents, organic solvents, etc. These chemical substances extract the fats from the skin and cause irritations and dermatitis. Sometimes, they penetrate deeply and mix into the blood. The gastrointestinal method of entrance of toxins into the organism occurs when the toxins enter the organism orally, sublingually or rectally.

Absorption

The entrance of the toxin into the blood circulation system by passing through the biological membrane is defined as "absorption." Toxins are absorbed in different tissues according to their means of entrance. The toxic effect of the toxins begins at the point of entry, such as the skin, oral region or inhalation membranes. These membranes contain a thick layer, as in the skin, gastrointestinal system or living protoplasm. It is supposed that the membranes have small holes on their surfaces. These holes permit the passage of water and any particles having a molecular weight equal to 100 or less. Some special membranes, such as those found in the kidneys. have larger holes. Toxic substances pass through the membranes by four basic mechanisms:

1. **Simple Diffusion:** This mechanism is considered to be the most important means of absorption of the toxic material. Substances that are soluble in lipids pass through the membranes more easily and diffuse in the aqueous phase on the other side of the membrane. Ionized compounds hardly diffuse through the membrane at all, due to their reaction with the lipids and proteins of the membrane.
2. **Filtration:** The small molecules (molecular weight <200) dissolved in the aqueous phase that fills the holes of the membrane, can pass through these holes. Defined as filtration, this occurs with the aid of osmotic or hydrostatic pressures. The holes of the membrane differ according to different organs of the body.
3. **Active transport:** Substances that are easily soluble in lipids and have high molecular weights cannot pass the cell membranes by simple diffusion or filtration, and it is suggested that they pass the cell membranes by means of a special mechanism. This is a very complex mechanism that is important in the elimination of toxic materials, rather than their absorption.

4. **Endocytosis:** The extensions that extend from the cell membrane wrap toxic materials and draw them into the cell. High molecular weight and colloidal particles can pass the cell membranes by this method. The dislocation or exit of the high molecules and particles from the cell by splitting of the cell is called *exocytosis*.

Enterohepatic Cycle

Chemical substances are transferred into the lymph system or into the liver after they are absorbed in the intestines, where they are partially or totally biotransformed and discharged into the bile. Therefore, toxic materials circulate by beginning with the intestines, passing to the liver, then, via the back to the intestines, causing the toxic material to be present in the blood for a long time. This process, called the "enterohepatic cycle," continues until the excretion of the chemical substance by other mechanisms in the body. It was found that substances like DDT, aldrin and dieldrin, which are in the form of chlorine hydrocarbon insecticides, form an enterohepatic cycle in experimental animals. This type of cycle causes an increase of toxic substances in the liver.

Special Biological Inhibitors

Some systems in the organism block the distribution of foreign chemical substances and inhibit their harmful effects to organs that have important functions in the body. The *blood–brain barrier* is a system with very low permeability to the toxic substances, which means that the entrance of many toxins into the central nervous systems is inhibited. Because this barrier is not completely developed in newly born babies, they are more susceptible to toxic substances than adults. Substances like mercury, lead, aluminum and alcohol pass this barrier and damage the central nervous system. Alcohol also increases the permeability of the blood–brain barrier.

The *placental barrier* inhibits the entrance of toxic substances to the fetus in pregnant women. However, some substances, such as radiation, can pass this barrier. Lipid-soluble xenobiotics cross the placenta by diffusion and are deposited in the fetus. Organophosphorus insecticides such as methamidophos and viruses such as rubella can also cross the placenta and cause teratogenic effects.

DISTRIBUTION, ACCUMULATION AND TRANSLOCATION

Distribution

A chemical substance, after being absorbed in the plasma, is distributed throughout the body. Although this distribution takes place rapidly, the ability of chemical substances to pass through the cell membrane changes according to their affinity to the components of the body. Some toxins cannot pass the cell wall, so they have limited distribution. Water-soluble substances with low molecular weights (<50) are distributed in the body fluids and pass through the membrane easily. Lipid-soluble substances pass through the membranes by simple diffusion, while big polar molecules can pass through the membranes only by special transport mechanisms. The role of vascular fluid is important in the distribution of toxic material because human plasma constitutes 4% of the total body weight and 53% of blood volume. Thirteen percent of the body weight is interstitial fluid, while 41% is intracellular fluid. Toxic substances may be distributed only by way of plasma and, in this way, it may reach a high concentration in the vascular tissue. Transport through the lymphatic system is not important. Transport within the plasma is binding of the toxic substances with plasma proteins, especially with albumins, followed by globulins. Both anions and cations bind to albumins. Binding to plasma proteins makes the chemicals biologically inactive and less available for filtering by the kidney, and, when the plasma proteins become saturated, further absorption of toxicants may result in toxicity. Although many toxic substances are "inert" from the point of view of their chemical properties, they may bind to some biological molecules reversibly.

Generally, toxic substances are bound to proteins and other biomolecules as follows:

$$\text{Toxic substance (free)} \rightarrow \text{protein bounded toxic substance}$$

The toxic substance that binds to the plasma proteins moves in the circulatory system until it is bounded to another biomolecule or to a tissue component.

Accumulation

Toxic substances, according to their different properties, accumulate in several parts of the body. This accumulation may occur as a result of the binding of the toxic material specifically in a definite region or by being soluble in lipids. If the toxic substance does not show biological effect in a certain place, this place of accumulation is called the "place of storage." In order to produce a toxic effect, the substance must leave the place where it has accumulated and reach its target point. For example, lead accumulates in bones and is inactive there; it shows its toxic effect as free lead ions (Pb^{++}) in the blood. Cellular binding results from the affinity of a substance to some cellular components. Toxins are accumulated in different cells, tissues or organs of the body as follows:

Accumulation in the organs: The liver and kidneys bind toxins more easily than the other organs. It is thought that these organs bind toxins to proteins or bind them by active transport in the blood system. It was shown that a protein (Y.protein or ligandin) found in the cytoplasm of the liver has a great affinity for organic acids. This protein also binds the carcinogenic azodyes. Another protein in the liver and in the kidneys binds cadmium.

Accumulation in lipid tissues: Lipid-soluble substances such as chlorine hydrocarbon insecticides (DDT) and polychlorinated biphenyls (PCBs) accumulate in lipid tissues and show their toxic effects in the blood. In overweight people, these substances pass into the blood during a high level of energy consumption by the fatty tissues and can cause serious toxic effects.

Accumulation in the bones: Many toxic chemicals, such as fluorides, lead, strontium, radium and tetracyclines, accumulate in the bones. Bones are places of storage only for some chemicals such as lead. However, fluorine shows its toxic effect in the bones by substitution with Ca^{++} ions. Radium also has similar toxic effects in the bones.

Accumulation in other tissues: Toxic materials may accumulate in different tissues according to their different properties. For example, arsenic accumulates in keratin-rich tissues, while iodine collects in the thyroid glands and carbon monoxide accumulates in the hemoglobin of the blood.

Translocation

The storage places of toxic materials may change through time. The first place of accumulation is related to a high degree of blood flow, to the high permeability of the tissue to the toxic substance and to reaching a binding place that is suitable for the toxic substance (such as a protein, a tissue component or another suitable biomolecule) to accumulate. When the chemical substance moves to another place where there is less blood flow, but is more suitable for accumulation, this transport of the toxic substance is named translocation. The accumulation and distribution of lead is a good example. In the first step of absorption, the important part of lead accumulates in the liver and kidneys by erythrosine. Within 2 hours, half is accumulated in the liver. After 1 month, lead distributes and accumulates in bones by substituting the Ca^{++} ions. By this method, 90% of absorbed lead accumulates in the bones.

METABOLISM

Generally, metabolism is defined as the total chemical reactions that occur in the organism for the continuation of life. The chemical changes of chemical substances that are foreign to an organism are termed biotransformation. Biotransformation can also be defined as the metabolism of xenobiotics. The chemical substances that enter the organism through various routes react with other substances with the aid of enzymes and change into their metabolites. The biotransformation of xenobiotics is important to the organism in several ways. Lipid-soluble apolar substances change into more polar-active metabolites by enzymatic reactions and show their toxic effects. The chemicals that gain biological activity by biotransformation are generally named "prodrugs." The foreign chemical or the xenobiotic, whether it is biologically active or not before biotransformation, changes into its metabolites, which show different effects by different mechanisms in the body, and then become inactive by conjugation and excreted. This type of biotranformation, related to a decrease in toxicity, is named *detoxification*.

The biotranformation of xenobiotics occurs with the aid of complex enzymes that are mostly located in the liver. Biotransformation can also occur in other tissues such as the intestine, kidney, lung, brain and skin. The enzymatic reactions in biotransformation are of two types: phase 1 reactions involving oxidation, reduction and hydrolysis; and phase 2 reactions, consisting of conjugation or synthesis. The most important enzyme systems involved in the first are the cytochrome P-450, containing monooxygenase enzyme systems that convert foreign compounds to derivatives that can undergo phase 2 reactions. It should be noted that biotransformation of xenobiotics is not strictly related to detoxification, although the enzymes carrying out phase 1 and phase 2 reactions are often referred to as detoxification enzymes. In some cases, the metabolic products may be equally or more toxic than the parent compounds. Aflatoxin M_1, which is as toxic as its parent compound Aflatoxin B_1, can be given as an example of food toxicants.

Factors that Affect the Rate of Metabolism

The rate of metabolism of toxic compounds to less-toxic products, or less-toxic compounds to more-toxic products is of considerable importance concerning the overall toxicity observed on exposure of animals to these compounds. The metabolism rates of xenobiotics are affected by several factors that can be summarized as age; differences in sex, strain and species; nutritional status; enzyme induction and pathological situations.

Age: The activity of many enzymes is very low in newborn babies and, generally, the babies of mammals are sensitive to many chemicals. Research has determined that the oxidative metabolism of foreign compounds is markedly reduced in fetal and newborn mice, rats, guinea pigs and rabbits. The rate of metabolism increases through the activity of enzymes, which increases with growing age.

Sex differences: A marked difference in the response of female as compared with male rats to a number of toxic foreign compounds has been noted. For example, the organophosphate insecticide parathion is approximately twice as toxic to female than male rats. This decreased susceptibility of the male rat to parathion is related to greater activity of the hepatic cytochrome P-450 monooxygenase system, which metabolizes parathion to nontoxic metabolites.

Strain differences: Strain differences in the metabolism of foreign compounds are seen among various mouse strains, which vary in their ability to metabolize the food contaminant benzo(a)pyrene.

Species differences: The differences in metabolism among the species are mostly related to differences in the activity of a particular enzyme or enzymes among those species. For example, the arly-acetic acids are conjugated predominantly with glutamine in man, whereas conjugation with glycine is the major reaction in the rat.

Nutritional status: The activity of enzymes and enzyme systems involved in the metabolism of foreign compounds is markedly affected by the nutritional state of experimental animals. Thus, in animals suffering from deficiency in calcium, copper, selenium, vitamin C, or protein, there is a decreased ability of enzymes to catalyze the oxidative reactions that occur in metabolism. It was also detected that carbohydrate-rich diets decrease the enzyme activities in metabolism.

Pathological situations: Because the liver is an important organ in the metabolism of xenobiotics, a pathologically adverse reaction in liver tissues would affect the activity of the enzymes seriously. For example, during hepatitis, the microsomal oxidation capacity decreases, so it was detected that Aflatoxin B_1 is more effective as a carcinogen to an organism where hepatitis B virus is present. Insufficient kidney functions also lower the elimination of chemicals from the body and cause increases in toxicity.

Enzyme induction: The activity of the hepatic cytochrome P-450-monooxygenase system in various animal species, including man, can be markedly increased by exposure to a large number of chemicals such as pesticides and the additives used in the food industry, or by various other industrial chemicals. This process of increased activity of these enzyme systems on exposure to chemicals is referred to as enzyme induction. Whether the induction or inhibition of the cytochrome P-450-containing monooxygenase enzyme system leads to an increase or a decrease in the toxicity of a particular compound is dependent on the toxic properties of the compound in question. If the monooxygenase-catalyzed metabolism of the compound leads to the formation of less toxic compounds, enzyme induction will result in a decrease in toxicity. On the other hand, inhibition of the monooxygenase system may lead to increased toxicity due to the increased half-life of the parent compound.

EXCRETION

Various routes from the body excrete the toxins that are absorbed and distributed in the living organism. The kidney is a very important organ for the excretion of toxins and probably more chemicals are eliminated from the body by this route than any other. Many toxic materials change into their polar metabolites and are eliminated in these forms. Lungs eliminate toxins in gas and vapor phases. Bile and feces are also important routes in toxin excretion, and saliva, sweat, tears and milk are also secondary routes for excretion of toxins. Because milk is consumed as a human food, the toxicological significance of milk and dairy products should be well evaluated.

4

THE EFFECT MECHANISM OF TOXINS

Toxins affect living organisms generally by chemical or physical-chemical ways and cause harmful results. These mechanisms can be examined in the following three groups:

1. Selective chemical effects
2. Nonspecific chemical effects
3. Special toxic effects

SELECTIVE EFFECTS

Chemical substances show their severe toxic effects under their lethal dose at definite concentrations in a region of the body specific for them. Normal cell components or cell membranes where the chemical substances show their effects in the living organism are named **target points**. These points may be very important for the cell's functions and the inhibition of these functions by the effect of the chemicals (toxins) damages the cell.

NONSPECIFIC TOXIC EFFECTS

Some chemical substances do not show their toxic effects at specific organs selectively, however they have a widespread effect. For example, strong acids and bases destroy all living cells, an effect probably caused by the denaturation and precipitation of the proteins in the cell membranes. Organic solvents are harmful to lipid membranes and cause the dissociation of nucleoprotein complexes. Many inert materials show these kinds of nonspecific effects, although they have different structures and do not involve functional groups.

SPECIAL TOXIC EFFECTS

Special toxic effects are classified into three groups: mutagenesis, carcinogenesis and teratogenesis.

Mutagenesis

The substances that form genetic differences in the genetic symbols of an organism are named genetic toxins. Many of the genetic diseases of humans have been shown to arise as a result of changes in the DNA structure, chromosomal structure or chromosome number. *Gene-locus mutations* or *point mutations* are changes in the DNA sequence within a gene. The process of forming gene-locus mutations is called mutagenesis or mutagenicity. The substances that cause mutagenesis are called *mutagens* and the species, which is exposed to mutagens, is called a *mutant*. Chemical substances that are considered mutagens are subdivided into three groups according to their effects on the DNA molecule. These are:

1. **Destructive:** Hydrogen peroxide, nitrates and nitrites can be given as examples of this group. Hydrogen peroxide causes genelocus mutations in microorganisms. Nitrates and nitrites are mutagenic toward bacteria, but no mutagenic activity has been detected in humans.
2. **Addition:** Chemicals that show this type of mutagenic effect are alkyl group-containing substances such as epoxides, dialkyl sulfates and lactones.
3. **Substitution:** Nucleic acids can be given as examples of the group of chemical substances that cause a change in the DNA molecule by substitution.

    It has been stated that mutagens can be produced during the cooking of foods. This was demonstrated in the broiling of dried fish and of beef, when amino acids and protein were pyrolized. Several mutagens have been identified as heterocyclic amines and some are carcinogenic.

Table 4.1 shows the mutagens and carcinogens formed in the gut and in food. As seen in the table, frying, grilling or broiling of protein-rich foods, especially meat derived from animal muscle tissue, in particular, causes the formation of extremely potent salmonella mutagens.

CARCINOGENESIS

Carcinogenesis or carcinogenicity is defined as the abnormal growth of somatic cells. The substances that cause carcinogenesis are called *carcin-*

Table 4.1   Mutagens and Carcinogens Formed in the Gut and in Food

Nitroso compounds (nitrosamines and nitrosamides)	Substances formed from reaction of amines with a nitrosating agent (nitrite or nitrate), e.g., N-nitroso-dimethylamine, -pyyrolidine, -piperidine
Mutagens formed during cooking	(i) Polycyclic hydrocarbons formed by pyrolysis of organic matter during grilling and smoking operations (benzo(a)pyrene)
	(ii) Pyrolysis products of proteins and amino acids formed during broiling of dried fish and meat (hamburger)
	(iii) Sugar caramelization products (ammonia caramel)
	(iv) Substances formed during Maillard reactions (5-hydroxymethylfurfural)
	(v) Fatty acid hydroperoxides and cholesterol epoxide from unsaturated fats
Ethyl carbamate	Found in wines treated with diethylpyrocarbonate and in naturally fermented foods and beverages

Adapted from Carr, 1985 and Larsen and Poulsen, 1987.

ogens. Historically, several types of chemicals were discovered to have carcinogenic potential in experimental systems after having first been suspected of causing cancer in man. Soot and coal tar were first suspected to be carcinogenic in the late 18th century, when Sir Percival Scott observed that many of his patients who had cancer of the scrotum were chimney sweeps. Exposure to soot began at a young age in Great Britain, where soft coal has been used for many centuries, and it was customary to train young boys to be chimney sweeps. Subsequent research has indicated that individuals are more sensitive to carcinogens when exposure first occurs early in life.

Carcinogens can be separated into eight classes and these classes can be divided into two general categories: *genotoxic* and *epigenetic* (Table 4.2).

Genotoxic carcinogens, because of their effects on genetic material, pose a clear quantitative hazard to humans. These carcinogens are occasionally effective after a single exposure and act in a cumulative manner. On the other hand, the carcinogenic effects of epigenetic carcinogens usually occur with high, sustained levels of exposure that lead to prolonged physiological abnormalities, hormonal imbalances or tissue injury.

Many animal and plant products have been identified to be carcinogenic to laboratory animals when administered for a suitable time and at an appropriate dose. Table 4.3 summarizes some compounds of natural origin in the human diet that are carcinogenic to experimental animals. Carcinogenic substances are also formed during the preparation of foods. For example, soil nitrate is a major source of plant nitrogen. It is converted to nitrite by plants and bacteria. A nitrite can react with a nitrogen-containing compound (amine) to form an N-nitroso compound that is a carcinogen. Secondary and tertiary amines in food and water are precursors for the nitrosatable amines. N-nitroso compounds, widespread in the environment, have been found in cured meats, a number of cheeses and certain dried dairy products (nonfat dry milk), some saltwater fish and some beers due to the barley malt. Recently, decreased industrial use of nitrates and nitrites as food additives and increased use of ascorbates have led to reduced N-nitrate levels in cured meats.

Some compounds that are added to foods intentionally or that contaminate foods during the food chain are shown in Table 4.4. The risks of all of these factors summarized in Tables 4.2–4.4 in the causes of human cancers are largely unknown. These substances have shown carcinogenic effects when tested on experimental animals for an appropriate duration and dosage. However, the sufficient period of exposure and dose necessary for carcinogenic effects on humans is a subject that still needs investigation.

TERATOGENESIS

Teratogenesis, or teratogenicity, is defined as giving birth to abnormally developed or underdeveloped babies. The fact that the environmental factors would cause congenital defects (malformations) was first stated by Murphy in 1929 by observing that some pregnant mothers who were exposed to x-rays bore mentally retarded children. In 1941, Gregg drew attention to the association of death, blindness and deafness among the offspring of women exposed to rubella (German measles) during pregnancy. Some 20 years later, the occurrence of 10,000 malformed infants born of mothers who had taken the drug thalidomide during their pregnancy was reported. The term *congenital defect* refers to all morphological, biochemical and functional abnormalities produced before or after birth. The substances that cause congenital defects are called *teratogens*.

Table 4.5 summarizes some selected substances that have been found to show teratogenic activity in animal experiments. The deficiency of some nutrients, as well as their consumption in excess amounts, may cause teratogenic effects. Vitamin A and nicotinic acid can be given as examples of these types of nutritive compounds. Teratogenic effects of contaminants in foods like nitrosamines, lead and aflatoxins, or some natural products like caffeine were also detected in animal experiments.

Table 4.2 Classes of Carcinogenic Chemicals

Type	Mode of Action	Example
Genotoxic Carcinogens		
Primary carcinogens	Form covalent bonds with macromolecules, interact with DNA	bis (chloromethyl) ether, ethylene imine, dimethyl sulfate
Secondary carcinogens	Gain electrophilic properties by enzymatic reactions (converted to type 1 by metabolic activation)	2-naphthylamine, benzo(a)pyrene, aflatoxin B_1, dimethylnitrosamine, vinyl chloride
Inorganic carcinogens	Lead to changes in DNA	nickel, chromium, lead, zinc, iron
Epigenetic Carcinogens		
Solid-state carcinogens	Exact mechanism unknown	asbestos
Hormones	Alter endocrine system balance	estradiol, diethylstilbestrol
Immunosuppressors	Affect the immune system	azathioprine
Cocarcinogens	Enhance the effect of primary and secondary carcinogens when given at the same time	pyrene, catechol, sulfur dioxide, ethanol
Promoters	Enhance the effect of primary and secondary carcinogens when given subsequently	phenol, bile acids, saccharin

Adapted from Weisburger and Williams, 1980.

Table 4.3   Compounds of Natural Origin in the Human Diet that are Carcinogenic to Experimental Animals

Complete Carcinogens

Hydrazines — Found in edible mushrooms (Japan)
Safrole — Sassafras plant, black pepper (spice flavor)
Pyyrolizidine alkaloids — Herbs, herbal teas, occasionally in honey
Gossypol — Unrefined cottonseed oil
Estrogens — Wheat germ, unpolished rice
Bracken fern (Japan)
Methylazoxymethanol from cycasin (cycad plants)
Carrageenan — Red seaweeds
Tannins — Tea, wine, plants
Ethylcarbamate — Wine, beer, yogurt

Tumor Promoters

Phorbol esters — Herb teas
Lynghyatoxin A — Contaminant from edible algae

Carcinogens from Foods Containing Mold

Aflatoxins (aspergillus)
Sterigmatocystin (aspergillus, penicillium)
Patulin — Apple mold (aspergillus, penicillium)
Luteoskyrin (penicillium islandicum)

From Carr, 1985, with permission.

Table 4.4   Compounds Added Industrially to the Human Food Chain that can be Carcinogenic to Experimental Animals

Food Additives	*Examples*
Colorings	amaranth
Flavorings	safrole, oil of calamus
Preservatives	diethylpyrocarbonate, 8-hydroxyquinoline
Sweeteners	cyclamates, saccharin, dulcin
Contaminants	*Examples*
Hormones	diethylstilbestrol
Pesticides	DDT, dieldrin, ethylene dibromide, chlordane
Pollutants from soil and water	arsenic, heavy metals, asbestos (cement water pipes), vinyl chloride (PVC water pipes), polychlorinated biphenyls (industrial coolants and flame-retardants), TCDD (toxic impurity of herbicide 2,4,5-trichlorophenol)

From Carr, 1985, with permission.

Table 4.5   Selected Teratogens in Animal Models

Dietary deficiency: Vitamins A, D and E, ascorbic acid, riboflavin, thiamine, nicotinamide, folic acid, panthothenic acid, trace metals (Zn, Mn, Cd, Co), protein

Vitamin excess: Vitamin A, nicotinic acid

Carbohydrates: Galactose, 2-deoxyglucose, bacterial lipopolysaccharides

Antibiotics: Penicillin, tetracyclines, streptomycin, sulfanilamide

Chelating agents: EDTA

Trace metals: Methylmercury, inorganic mercury salts, lead, thallium, strontium, selenium

Azo dyes: Trypan blue, Evans blue, Niagara sky blue 6B

Agents producing hypoxia: Carbon monoxide, carbon dioxide

Chemicals: Nicotine, quinine, saponin, boric acid, salicylate, 2,3,4,8-tetracholorodibenzo-*p*-dioxin, nitrosamines, caffeine

Solvents: Dimethyl sulfoxide, chloroform, 1,1-di-chloroethane, carbon tetrachloride, benzene, xylene, cyclohexanone, propylene glycol, alkane, sulfonates, acetamides, formamides

Mycotoxins: Rubratoxin B, aflatoxin B_1, ochratoxin A

Physical agents: Radiation

Infections: 10 viruses known, including rubella

Adapted from Harbison, 1980.

Teratogenic events have not been experimentally observed in human beings, however, some hormones such as estrogens and cortisones are accepted as teratogenic to man. Also, the chemicals that show teratogenic effects in animals were found to be harmful according to the results of some clinical observations.

5

TOXICITY TESTS

The primary objective of toxicological testing is to determine the effects of chemicals on biological systems and to obtain data on the dose–response relations of the chemicals. These data may provide information on the degree of hazard to man and the environment associated with a potential exposure related to a specific use of this chemical. The major difference among toxicity tests is the dose employed and the length of exposure to the chemical agent. The term *dose* is used to specify the amount of chemical administered, usually expressed per unit body weight. *Effect* and *response* are often used interchangeably to denote a biological change associated with an exposure or dose, either in an individual or in a population. Some toxicologists have, however, found it useful to differentiate between an effect and a response by applying the term "effect" to a biological change and the term "response" to the proportion of a population that demonstrates a defined effect. In this terminology, response means the incidence rate of an effect. For example, the LD_{50} value can be described as the dose expected to cause a 50% response in a population tested for the lethal effect of a chemical.

The toxicological tests are generally applied by using experimental animals (*in vivo* test systems), short-term (*in vitro*) tests, epidemiological studies and biomarker research.

IN VIVO TEST SYSTEMS

The general requirements for conducting *in vivo* toxicity tests can be summarized as follows:

- Groups of healthy animals, housed under suitable conditions, are exposed to the graded doses of the test chemical.

- Rats, mice, guinea pigs, rabbits and hamsters are mainly used. In some cases, dogs, swine, nonhuman primates or other species are also used.
- A control group is given the dosing vehicle or is sham treated. Laboratory tests are conducted on treated and control animals. Detailed records are maintained for each animal.
- Following completion of the test, all animals, including controls, are subjected to a pathological examination and data is analyzed by statistical procedures.
- Animals should be genetically stable and adequately defined and identified as to colony source. The controls and treated animals should be of the same strain and species, age, weight range and supplied from the same source. The diet fed to animals should meet all nutritional requirements and should be free of toxic chemical impurities that might influence the outcome of the test.

In general, *in vivo* toxicity tests can be classified as acute, subacute/subchronic and chronic toxicity tests.

Acute Toxicity Tests

Acute toxicity is defined as adverse effects occurring by a short-time administration of a single dose or multiple doses given within 24 hours. The most frequently used acute toxicity test is determination of the median lethal dose (LD_{50}) of the compound. The LD_{50} has been defined as a statistically derived expression of a single dose of a material that can be expected to kill half of the experiment animals. A lethal dose of hydrogen cyanide (50–60 mg) induces death within minutes. Cicutoxin, the principal toxin of hemlock, kills so rapidly that cattle often die before the water hemlock they have eaten passes beyond the esophageal groove. Death that occurs after the first 24 hours is more likely to be due to delayed toxic effects. Signs occurring after the first 24-hour period may give some indication of the effect that the chemical may have at lower levels when administered for longer time periods.

The mouse, rat and dog are the most commonly used species for acute toxicity testing. Both the rat and the mouse should be used, as marked differences in the LD_{50} between these two species are not common. Usually 8–10 rodents (4–6 animals of each sex) are used per dose group.

In the method of administration, the route by which man would be exposed is used. If the route is oral, the compound should be administered by gavage (fed directly into the stomach by means of a tube) rather than mixed into the diet. In some cases, the administration of the chemical along with food has been shown to increase its toxicity but, in general, the oral toxicity of the compound is greatest when it is administered by gavage to animals that have fasted.

In postmortem examination, all animals dying during the observation period and all surviving animals should be autopsied by a qualified pathologist. The autopsy should include gross and histopathological examination of all organs.

Subacute/Subchronic Toxicity Tests

Acute toxicity and the LD_{50} in particular are not relevant for food additives and are of limited value in relation to industrial or agricultural exposure to pesticides. As far as consumer exposure is concerned, long-term effects of repeated exposure to subacutely toxic doses gain importance in food toxicology.

The subchronic toxicity test involves daily or frequent exposure to the compound over a period of up to 90 days or 3 months. In rats, subchronic toxicity studies would take 90 days, whereas 14-, 21- and 28-day studies are generally referred to as subacute. These studies should be conducted on two species, rodents and nonrodents. Traditionally, the rat and the dog are selected because of their availability and the large amount of background information available on them. When rats are used, the test should be initiated just after weaning so that the observation can be made during the period of most rapid growth. At least 10 animals of each sex should be included in each dose group and the experiment should continue for 10% of the animals' lifetime or about 3 months. If it is desired to study the pathogenesis and reversibility of induced lesions or biochemokinetics, it is recommended that observations be made at 3-week intervals during exposure and should be continued up to 3 months following termination of exposure. In food toxicology, subacute/subchronic toxicity tests are often used to define the NOAEL (no observable adverse effect level) values for noncarcinogens and the MTD (maximum tolerated dose), which is the highest level of a food substance that can be fed to an animal without inducing obvious signs of toxicity other than those due to cancer.

Chronic Toxicity Tests

Chronic toxicity refers to an adverse effect that requires some time to develop. This test is usually conducted with the aim of establishing NOAEL values to be used for setting ADI (acceptable daily intake) values, tolerance limits of chemical additives and contaminants in food toxicology. A wide variety of animal species is used, but mostly rodents are selected. Larger animals (dogs and monkey) are also used for large samples of blood. The chemical is administered daily over the entire treatment period and covers the major portion of the life span, beginning from early exposure in life.

Carcinogenicity and mutagenicity tests are applied by using rodents — specifically, mouse, rat and Syrian golden hamster. Rodents are preferred over other species because of their susceptibility to tumor induction, their relatively short life span, the limited cost of their maintenance, their widespread use in pharmacological and toxicological studies, the availability of inbred strains and the large body of information available on their physiology and pathology. Nonrodents, in particular dogs and primates, can also be used. However, the high cost of maintenance, the long period of observation and the impracticability of using insufficiently large numbers are some disadvantages of nonrodents in long-term bioassays. In tests for carcinogenicity of the chemical, the amount of carcinogen required to induce cancer in half of a group of exposed animals is referred to as the tumor dose (TD_{50}).

As in testing for acute toxicity, the route of administration in chronic and subacute toxicity tests should be that through which man is likely to be exposed. For gases and volatile industrial solvents, inhalation studies are recommended, while, for food additives, pesticides and contaminants ingested by consumption of food or water, the oral route is recommended. Incorporation of the test chemical into the diet or drinking water is an appropriate means of administration, but care must be taken to ensure the stability of the chemical in the dosing medium. In some cases, the chemical may be unpalatable and administration by gavage, or, in the case of dogs, by capsules, may be necessary. Biochemical organ function tests, physiological measurements and metabolic studies are conducted and hematological information is taken after application of chronic and subacute toxicity tests.

SHORT-TERM (*IN VITRO*) TESTS FOR DETECTING MUTAGENICITY AND CARCINOGENICITY

Short-term *in vitro*, or test-tube studies are also performed to determine the mutagenic or carcinogenic potential of a product. The Ames test is designed to check the mutagenicity and the alterations in the genetic material. This test, done with bacteria, can rapidly demonstrate whether a chemical can cause a certain type of genetic damage. However, the results in a simple test system are not easily transferable to multicelled organisms, especially to humans. Thus, the results taken from this test are not useful in establishing the dose at which a particular chemical may cause genetic damage in humans and they cannot be applied to long-term exposures.

In vitro tests are also used as indicators to determine the carcinogenic potential of chemicals because there is a relationship between genetic damage and cancer. Thus, the chemicals that test positive in short-term assay are considered to be good candidates for long-term animal studies. The evaluations performed by considering the carcinogenicity and genotoxicity data have a significant role in interpretation of the cancer studies. However, there is a need for standardization of test protocols and validation of the results against traditional methods in order to make *in vitro* tests acceptable as reasonable indicators of carcinogenicity. It is stated that *in vitro* tests evaluate only one type of toxicity, i.e., genotoxicity, thus when a single genotoxicity test is used to determine carcinogenicity, several tests detecting different endpoints (point mutations, clastogenicity) should be performed. In food toxicology, certain food additives, such as sulfur dioxide, are mutagenic *in vitro* but neither mutagenic nor carcinogenic *in vivo*. So, there is a need for long-term carcinogenicity studies besides the short-term *in vitro* tests, to establish the values such as NOAEL or MTD for these compounds. Although several alternative *in vitro* methods have been suggested in the areas of prenatal toxicity (e.g. *in vitro* culture of mammalian embryos), tumor promotion (e.g., loss of cell–cell communication) and organ toxicity (e.g., cultured organs), these methods give reproducible results when they are interpreted by animal experiments.

HUMAN STUDIES

Because extrapolation of data from animals to humans poses some problems because of the differences between the animal and human species, as well as the individual variations among the same strain, human studies are also conducted for determining the toxicities of different chemicals. Human studies can be conducted by performing epidemiological studies and biomarker research.

Epidemiological Studies

Epidemiology is the study of human populations to establish correlations among the particular environmental conditions and specific health effects. Epidemiological studies are performed to evaluate the relationships among exposures to particular toxic substances and specific chronic effects. Some examples of epidemiological studies are those linking substances like food colors such as tartrazine with allergic reactions in some people, aflatoxin B_1 with liver cancer, and prevention of lung cancer by consumption of β-carotene. The major disadvantages of epidemiological studies is the wide variety of substances to which humans are exposed, which causes difficulty in determining what types of effects a specific compound will cause.

Population variability also affects the outcomes of epidemiological studies. In case-control or the retrospective type of studies, two populations are mainly studied. The first population consists of individuals who demonstrate the toxic effect of interest and the second is made up of those who do not. The two populations are matched by considering all other variables such as age, socioeconomic status, medical history and, if they are working, the conditions of the workplace. The past histories of exposure of the two populations are investigated to observe differences that can be related to the toxic effects of the compound of interest. In the prospective or cohort type of epidemiological study, a population is followed from a set time into the future. These types of studies may take years to be completed, complicated by the fact that the individuals taking part in the study may behave differently, such as having different habits of preparing or consuming foods. Although epidemiological studies have such limitations, they can be used to confirm the results of animal experiments. A chemical substance that has been found to produce a specific toxic effect on animals can be examined in the human population, and the results of epidemiological studies can provide additional evidence for determining toxicity of the examined substances. An example that can be taken from food toxicology regarding the usefulness of epidemiological studies is the correlation of the development of vaginal cancers in the daughters of women who took the hormone diethylstilbestrol (DES) during pregnancy.

Biomarker Research

A biomarker is defined as a parameter at the biochemical, physiological, enzymatic, morphological or cellular level that reflects exposure or the pathophysiological status of the individual. Biomarkers reflect some phase between external exposure and eventual effect (disease) and include factors that may modify transition states between those phases (individual susceptibility, nutrition). The biomarker concept covers the entire chain between primary exposure and the effect that occurs, such as tumor formation, bone fracture, hair loss, etc. By measuring well-defined and selected biomarkers, it is possible to study potentially harmful or beneficial effects of compounds in the target species itself — the human being. A framework for biomarkers depicting the modulating roles of nutrition and susceptibility is shown in Figure 5.1. The application of biomarkers has also been used in food toxicology. The biomarkers of internal dose that can be used include analyses of nutrients or compounds or metabolites in blood, plasma, blood cells, feces, urine, milk, bile, saliva, vomitus, expiratory air, sputum cells, sweat, skin exfoliations, hair, etc.

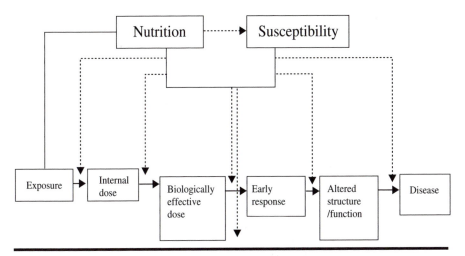

Figure 5.1 Framework for biomarkers depicting the modulating roles of nutrition and susceptibility. From Hermus et al., 1994, with permission of S. Karger AG.

Toxicity Tests for Hazard Characterization in Foods

Several types of hazards may have toxicological significance in food consumption. These hazards are present in foods because of substances such as unintentional xenobiotic (any harmful element that is foreign to the living organism) contaminants or intentional xenobiotic food additives. The toxicity methods described in this chapter can be used for evaluation of these hazardous substances. Table 5.1 summarizes the toxicological methods that can be used to discover the presence of xenobiotic compounds in foods.

Table 5.1   Routine Toxicity Tests Commonly Required in Determining the Degree of Hazard of Xenobiotic Compounds In Foods

Toxicity Test	Outputs
Acute toxicity (Single dose)	Nature of acute (overdose) effects Target organ(s) or systems Median lethal dose (LD_{50})
Subacute or Subchronic toxicity (28–90 day repeat dose studies, usually in diet or drinking water)	Nature of toxicity; target organ(s) or systems Dose–response characteristics No observed (adverse) effect level (NOAEL) Maximum tolerated dose (MTD)
Genotoxicity (Mutagenicity or clastogenicity, *in vitro* in prokaryotes and eukaryotes, including mammalian cells; short-term *in vivo*)	Evidence of potential genotoxicity (±?) Potency
Chronic toxicity* (Long-term dietary administration, e.g., 6 months to 2 years in rats and mice)	Nature of chronic toxicity; target organ(s) or systems Cumulative effects; dose–response relations NOAEL
Carcinogenicity* (Long-term administration at the MTD, e.g., 2 years in rats and mice)	Carcinogenic potential (±?) Potency
Reproductive toxicity (Single or multiple dose studies during pregnancy; multigenerational studies with dietary administration prior to and during mating, gestation and suckling)	Effects on fertility (male and female); fetotoxicity; teratogenic potential; effects on lactation and postnatal development Dose–response characteristics for these effects
Metabolism and toxicokinetics (Comparative absorption, distribution, metabolism and excretion in test species and human)	C_{max}; AUC, bioavailability. Dose-dependent kinetics, saturation of detoxification mechanisms Species-independent metabolism

* Chronic toxicity and carcinogenicity studies can be combined

From Walker, 1999, with permission.

6

NATURAL SOURCES OF TOXICANTS IN FOODS

All chemicals, whether synthetic or naturally occurring, exhibit toxicity at some level of exposure. Even drinking excessive amounts of pure water can kill through the induction of an electrolyte imbalance. Also, there is a condition called "water toxicity" that is due to decreased renal capacity to excrete water. In foods that are normal constituents of our diets, some natural chemicals can be present in doses that are sufficient to produce harmful effects. These chemicals may cause toxic reactions when eaten in normal amounts by a person who tends to misuse a certain food or consume it in large amounts. An example of this kind of a situation is consuming a food that is rich in protease inhibitors without the application of necessary cooking procedures. An abnormal food constituent eaten in normal amounts may also cause toxic reactions. An example of this situation occurs in honey, when the bee collects nectar from plants containing poisonous (even carcinogenic) alkaloids, and acetylandromedel or andromedol is transferred to the honey. A normal constituent of a food such as lactose in milk may be toxic (lactose intolerance) to an individual whose body lacks the enzyme lactase. Nitrates that are found in spinach and other green leafy vegetables can also cause toxic reactions by combining with the amines in the food's structure to form carcinogenic nitrosamines. Table 6.1 summarizes the adverse reactions that can occur through the consumption of natural foods.

The different types of constituents in natural foods that may be toxic to different individuals, as described above, can be classified and described in the following groups.

Table 6.1 Adverse Reactions: Causes and Foods Commonly Involved

Type of Reaction	Cause	Food/Food Compounds Involved
Food allergy	Allergic reaction to a food caused by an over-sensitive reaction of the body's immune system	eggs, milk, wheat, fish, shellfish, nuts, peanuts, soybean, rice
Pharmacological	Absorption of certain amines from foods containing high amounts	fermented foods (cheese, red wine, sauerkraut, fermented sausages), fish products
	Other substances with pharmacological type of action	caffeine
Enzyme defects	Failure of normal enzymatic breakdown after absorption	alcohol, fructose, amines
	Failure to digest, so that unabsorbed substances reach the lower intestine, where they are fermented	lactose (in milk), fats
Irritant	Often made worse by acid which refluxes from the top end of the stomach to cause heartburn	strong spices and flavors, sulfites
Toxic	Toxins	some shellfish, badly stored food (e.g., green potatoes), some fungi
Psychological food intolerance	Emotional reaction to a food (reaction does not occur if food is not recognized)	

From Lessof, 1998, with permission of International Life Sciences Institute.

ALLERGENS

The term allergy is generally used to describe an altered immunological reactivity to a foreign material, which is then called an allergen. Allergens are usually normal food constituents and the abnormality rests in the individual who has an altered reactivity to such substances. The intensity of the reaction depends on the degree of hypersensitivity of the person consuming the food rather than on the quantity consumed. Food allergy involves the reaction of food protein antibodies to the

Table 6.2 Symptoms of Food Allergy

Gastrointestinal Symptoms

Abdominal cramps
Bloating
Colic
Diarrhea
Nausea
Vomiting

Respiratory Symptoms

Asthma
Laryngeal edema
Recurrent cough
Rhinitis
Sneezing
Wheezing

Dermatological Symptoms

Angioedema
Eczema
Erythema (skin inflammation)
Pruritis
Urticaria

Other

Anaphylaxis

From Lessof, 1998, with permission of International Life Sciences Institute.

immune system. Accompanying symptoms are the result of this inappropriate immune response. These symptoms can occur very quickly, often within minutes, and hence are termed "immediate" reactions. In the oral allergy syndrome, the lips, cheeks, tongue or mouth may itch or swell within minutes of contact with such foods as eggs, nuts or peanuts. A food allergy is usually detected by subjective association of symptoms to ingested foods and is treated by dietary restriction. The major symptoms of food allergy are shown in Table 6.2. Foods that are commonly involved in allergies are eggs, peanuts, cow's milk, soy, wheat, peas, fish, shellfish and nuts, but reactions can also occur to spices such as mustard, or to vegetables, for example, celery, tomato and eggplant. Some of the major allergens in foods have been identified and shown to be proteins. The foods that cause allergic symptoms in some individuals are given in Table 6.3. In addition to this list, food additives that are added to foods intentionally, especially artificial colorings, flavorings and gums may cause allergic reactions in some people.

Table 6.3   Foods that Cause Allergic Reactions in Some Individuals

Food	Examples
Cereal grains	wheat, corn, rye, barley, buckwheat (Baker's eczema, Baker's asthma)
Vegetables	carrots, potatoes, squash, celery
Legumes	peanuts, lentils, soybeans
Fruits	strawberries, bananas, pineapples, mangos, tomatoes
Animal foods	cow's milk, eggs, crustacea, mollusks
Nuts, seeds and beans	cottonseed, castor beans, mustard seed, flaxseed
Beverages	chocolate, coffee, chicory, tea, carbonated beverages that contain artificial colorings or flavorings, beer, wine

Adapted from Perlman, 1980 and Taylor, 1990.

ALKALOIDS

Alkaloids, which frequently have pharmacological properties, are bitter components in plants. Solanine is an alkaloid in potato that acts like a natural pesticide against common potato pests. It is usually high in sunburned (green under the skin) or blighted potatoes. It cannot be washed away or decreased by cooking. Cooked potatoes that contain solanine have a bitter taste and cause a burning sensation in the throat. Tomatine is an alkaloid that is present in tomatoes as a natural pesticide. Solanine and tomatine are naturally beneficial to the plants, but not for the mammalian metabolism. Caffeine and theobromine are alkaloids found in coffee, tea and chocolate. Caffeine inhibits DNA repair under some conditions, and heavy coffee drinking has been implicated (but not proven) in epidemiological studies of cancer of the ovary, bladder, pancreas and large bowel. Theobromine, at high levels, retards its own breakdown and, when taken as chocolate, does not present a significant problem because of limited bioavailability in this form. Pyrrolizidine alkaloids are chemicals found in as many as 6000 plant species or 3% of flowering plants. Plants containing these compounds are distributed in all climatic regions of the world and are classifed as carcinogens in foods of natural origin. The major organ affected by the pyrrolizidine alkaloids is the liver, however, damage to the lungs, heart and kidney has also been reported. Plants containing these alkaloids may grow as weeds in food crops such as wheat or corn and may be harvested with the grain. Quinine, an alkaloid that is used as a drug, is found in the beverage tonic water as a food additive. Some of the side effects include skin rash, tinnitus, slight deafness, vertigo and slight mental depression.

Table 6.4   Cyanide Levels of Various Plants

Food	HCN Yield (mg/100g)
Apricot pit	60
Peach pit	160
Bitter cassava	3–245
Sorghum (leaves and young shoots)	60–240
Bitter almonds	250–290
Lima bean varieties	17–312

Adapted from Taylor and Schantz, 1990; Jones, 1992.

CYANOGENS

Cyanogens are compounds that yield hydrogen cyanide when acted upon by stomach acids or certain plants or enzymes. As little as 30–250 mg is lethal to the adult male. Cyanogen ingestion can give rise to chronic neurological disease in man. This is related to tropical ataxic neuropathy and tropical amblyopia, which is seen in Nigeria and Tanzania, with cassava as the source. Cyanide has also been implicated in a type of diabetes in which cyanide levels lead to loss of β-cell function of the pancreas. The amounts of cyanide released from some common plant tissues are given in Table 6.4.

As can be seen from the table, the highest cyanide yield is common in bitter almond and lima bean varieties. Among the lima beans, colored Java beans (312 mg/100 g) and Puerto Rico black beans (300 mg/100g) are the highest cyanide sources. Cyanide shows its toxic action by preventing oxygen binding to hemoglobin, which causes "cyanosis," a bluish coloration of the skin and mucous membranes. Although cooking destroys the cyanide-releasing enzyme in the plant (emulsin), acids in the stomach still can release the cyanide. The cyanogenic glycoside found in almonds, bamboo shoots, sorghum, chokecherries and wild black cherries, and in the pits of apple, apricot, cherry, plum, quince and peach are dhurrin (so-called bitter almonds) and amygdaline (laetrile). It has been stated that just 12 bitter almonds can kill a child. In some countries, marzipan and almond paste have been a source of cyanide. Myristicin is a potent hallucinogenic chemical produced by a variety of common plants. It has been shown to be present in dill, celery, parsley, parsnip and mint, but is most widely consumed by humans as a component of nutmeg. Consumption of as little as 500 mg of nutmeg may produce psychoactive symptoms, while doses of 5 to 15 g of powdered nutmeg may result in euphoria, hallucinations and a dreamlike feeling that may be followed by abdominal pain, depression and stupor.

ERUCIC ACID

Erucic acid is *cis*-13-decosenoic acid, that is, a 22-C monounsaturated fatty acid that is the principal fatty acid in the triglycerides of rapeseed oil and other seed oils of the *Brassica* genus. It retards growth, and has been implicated as liver degeneration and kidney nephrosis. The feeding of rapeseed oil caused fat accumulation (lipidosis) in the heart and skeletal muscles of rats. In the heart, the deposition of these fat droplets was related to the level of erucic acid in the feed. The consumption of the feed for a period of about 16 weeks caused development of physiological changes in the heart of the test animals. Lesions due to muscle cell destruction, inflammation and scarring of tissue were observed. Other docosenoic acids such as cetoloic acid (*cis*-11-docosenoic acid), found in fish and marine animal oils and in partially hydrogenated fish oils used in blended human dietary fats, have been shown to cause lipidosis when fed at high concentration to test animals, but to a significantly smaller extent than erucic acid. As little as 15% erucic acid can cause myocarditis, which is a severe health problem. Plant breeding has developed a low-erucic-acid species of rapeseed called canola, which can be used as cooking oil or processed into margarine.

FAVISM

Fava beans (*vicia fava*) or broad beans, widely eaten in England and countries around the Mediterranean, are associated with the disease named **Favism.** When ingested by persons who have been previously sensitized either by eating pollen of the blossoms of the fava plant or by prior ingestion of the beans, severe illness may occur. Favism is an inherited, sex-linked metabolic disturbance occurring in some people of Mediterranean and Asiatic origin. It is also seen in parts of Africa and among African-Americans. It appears to be a hereditary deficiency of the enzyme glucose-6-phosphate dehydrogenase (G6PD). Illness appears within 1 hour of ingestion and the major symptoms are dizziness, vomiting, diarrhea and severe prostration. Eating of fava beans by genetically sensitive individuals results in hemolytic anemia and hematuria (blood in the urine), which is caused by the rupturing of the older blood cells by the fava bean nucleosides vicine and covicine. Children are especially susceptible, and favism is occasionally fatal. Due to high levels of monoamine oxidase inhibitors, which cause headaches, palpitations and sharp increases in blood pressure, fava beans can also affect individuals who do not have favism. Treatment is symptomatic, with replacement of electrolytes and blood transfusion in severe cases.

FUNGI

A number of poisonous mushrooms grow from the spring to autumn in temperate-climate woods, by the roadside and even on cultivated land. The toxicological events that can result from the consumption of mushrooms range in severity from mild to fatal and, in latency period, from less than 2 hours to greater than 72 hours. Mushrooms can be grouped into different categories, depending upon their structures and their symptoms.

Amatoxins are the most dangerous group of the mushroom toxins. *Amanita phalloides*, also known as "death cap," contains 2–3 mg of amatoxins per gram of dry tissue. A single mushroom of this kind can kill an adult human. Poisoning with *A. phalloides* is caused by alpha, beta and gamma amanitine, a heat-stable peptide that cannot be destroyed by cooking. Symptoms of amatoxin poisoning begin within 6–24 hours of ingestion of the mushrooms. In the first stage, abdominal pain, nausea, vomiting, diarrhea and hyperglycemia occur. The second stage is a short period of remission and, in the third stage, severe liver and kidney disfunction occurs, with symptoms including abdominal pain, jaundice, renal failure, hypoglycemia, convulsions, coma and death. Death results from hypoglycemic shock, usually between the 4th and 7th days after the onset of symptoms. Recovery, with intensive medical intervention, may require 2 weeks.

Hydrazines, such as gyromitrin, cause a bloated feeling, nausea, vomiting, watery or bloody diarrhea, abdominal pains, muscle cramps, faintness and loss of motor coordination that typically occur 6–12 hours after consumption of *Gyromitra esculental* mushrooms. In rare cases, the illness can progress to convulsions, coma and death.

Muscarine (*A. muscaria*), is an alkaloid that acts like pilocarpine on the smooth muscles of the pupils and on the cells of external secretion-glands of the body. Within a few minutes to a few hours of consumption of mushrooms containing these toxins, the patient experiences perspiration, salivation and lacrimation (PSL) syndrome, blurred vision, abdominal cramps, watery diarrhea, constriction of the pupils, hypotension and a slowed pulse. Death does not usually occur when these are the only toxins in the poisonous mushrooms (e.g., *Clitocybe dealbata*).

Coprine is a toxin that causes symptoms only in conjunction with alcohol. Symptoms begin about 30 minutes after consuming alcohol and may occur for as long as 5 days after mushroom ingestion. The symptoms include flushing of the face and neck, distension of the veins in the neck, swelling and tingling of the hands, metallic taste, tachycardia and hypotension, progressing to nausea and vomiting.

Certain groups of toxins contain compounds that affect the central nervous system and cause hallucinations. Isoxazoles, including ibotenic acid and muscimol, cause dizziness, lack of coordination, staggering, muscular jerking and spasms, hyperkinetic activity, a coma-like sleep and hallucinations within 30 minutes to 2 hours of ingestion. Indoles such as psilocybin and psilocin cause euphoric or apprehensive moods, unmotivated laughter and hilarity, compulsive movements, muscle weakness, drowsiness, hallucinations and, finally, sleep. Recovery is spontaneous. Death, associated with mydriasis, hyperthermia, hallucinations, loss of consciousness and convulsions has been reported in small children.

GOITROGENS

These substances are responsible for the pungent flavors of horseradish and mustard, and contribute to the characteristic flavors of turnip and cabbage. In certain cruciferous plants, they are associated with endemic goiter-hypothyroidism, with an enlargement of the thyroid gland. The genus *Brassica*, which is the primary dietary source for goitrogens (glucosinolates) includes broccoli, Brussels sprouts, cabbage, cauliflower, horseradish, turnip, rapeseed and mustard. Carrot, peach, pear, radish and strawberry may also contain goitrogens. Milk may also be a source if one of the *Brassica* species, such as turnips, was used as a fodder. Cooking and freezing reduce the amount of goitrogens in milk or vegetables. A high intake of these vegetables with an inadequate intake of iodine could precipitate goiter.

GOSSYPOL

Gossypol is a polyphenolic compound found in cottonseed and unrefined cottonseed oil. Cottonseed kernels contain about 0.6% gossypol. It reacts with proteins, reducing their quality. Gossypol also inhibits the conversion of pepsinogen to pepsin and limits the bioavailability of iron. General symptoms of gossypol toxicity are depressed appetite and loss of body weight. Acute toxicity is low, but chronically, it can cause death. Selective breeding programs have developed glandless cottonseed, which is essentially gossypol-free because the polyphenol transport from the roots to the seeds is blocked. Because of their high protein content, the seeds are often converted to animal feed. However, feed from landed cottonseed has toxic effects on monogastric animals. It was also found that cottonseed plant cultivars that produce much less gossypol are more susceptible to mold growth and aflatoxins.

HEMAGLUTININS (LECTINS)

These are naturally occurring constituents that are mainly found in legume seeds, but may also be found in other parts of plants. They are present at high levels in all legumes and grain products. Castor beans contain one of the most toxic, lectin (ricin), which is a protein capable of agglutinating red blood cells, which means that these beans are not suitable as food. Lectins can destroy the epithelia of the gastrointestinal tract; interfere with cell mitosis; cause local hemorrhages; damage kidney, liver and heart and agglutinate red blood cells. Because cooking with moist heat can reduce the toxicity of these compounds, their use in human diets is little cause for concern except at high altitudes, where the boiling point is reduced, or in situations where heat transfer is terribly uneven.

LATHYROGENS

Lathyrism is a neurological disease that occurs in India, China, parts of Africa and areas around the Mediterranean, especially Spain. The incidence of this disease increases during periods of drought, when people are forced to eat sweet pea (*Lathyrus sativus*) and vetch, which contain lathyrogens. In lathyrism, skeletal and aortic abnormalities occur due to the altered metabolism of connective tissues. The nervous system is also affected because the metabolism of the neurotransmitter glutamic acid is impaired. Neurological effects are seen only in humans, and males are more susceptible than females.

MARINE ANIMALS

Seafood poisonings may occur due to the poisons found in certain fish. Although these types of poisonings were historically considered a public health problem largely in coastal fishing sites, they have also spread to inland areas as a result of modern transportation and shipping of seafood products. Toxicants may be produced by the fish itself, by the marine plankton or algae consumed by the fish, or by bacteria that contaminate fish products.

Brevitoxin poisoning occurs in humans after consumption of toxic shellfish, usually clams or oysters, and usually taken between the months of November and March (in the U.S.). The dinoflagellate *Ptychodiscus brevis* causes red tides, which are associated with massive fish kills and have been limited to the Gulf of Mexico and areas off the coast of Florida. Because the dinoflagellate is taken in through the gills, it is toxic to fish but not to shellfish. *P. brevis* produces neurotoxins named brevitoxins A, B and C. These toxins are heat stable and produce neurotoxic effects causing nausea, diarrhea and paresthesia within minutes after ingestion of contaminated shellfish. Symptoms of brevitoxin poisoning include tingling of the face, mouth and throat, inversion of hot-cold sensation, irregular heartbeat and pupil dilation 1–3 hours after consuming the toxin. Recovery occurs within 24 hours in most victims.

Ciguatera poisoning is a gastrointestinal-neurological and occasionally cardiovascular syndrome, which is common throughout the Caribbean and Indo-Pacific islands. This type of poisoning appears to follow the spatial and temporal pattern of the distribution of the photosynthetic dinoflagellate *Gambierdiscus toxicus*, which is consumed by smaller herbivorous fish. Species of fish implicated include barracudas, surgeon fish, jacks, groupers, sea bass, sharks, triggerfish, wrasses, parrotfish and, especially, red snappers and eels. Ciquatoxin is a colorless, lipid- and heat-soluble molecule that increases membrane permeability to sodium ions causing depolarization of nerves. In humans, signs and symptoms include tingling of the lips, tongue and throat, followed by numbness, nausea, vomiting, abdominal pain, diarrhea, bradycardia, dizziness, muscle and joint pain and, in severe cases, paresis of the legs. Although most afflicted victims usually recover within a few weeks, in some cases, death may occur due to cardiovascular collapse. Ciguatera poisons may occur during any season in the tropics, but tend to develop most often in fish between 35°C north and 35°C south latitude during the spring or fall. Fish acquire the toxin through the food chain and all susceptible species in a certain area usually become toxic. Affected fish exhibit no odor or visible difference in appearance; therefore the toxin cannot be detected easily. The internal organs of the fish, such as liver, intestines and gonads, as well as roe, are more toxic than the more commonly consumed muscle tissue. Prevention of ciguatera poisoning is difficult and the best approach is described as surveillance for cases in the fish-consuming population, verifying the diagnoses by animal testing and warning the public in case of a hazard.

Puffer fish (fugu) poisoning has been known to occur as far back as 2000–3000 BC in China and Japan. The toxin that occurs is called tetrodotoxin. It is found in puffer fish, which are found worldwide in both tropical and semitropical waters. It is also found in ocean sunfish, porcupine fish, blue-ringed octopus and certain amphibians. The tetrodotoxin is concentrated in the liver, gonads and certain other internal organs that must be removed before the fish is consumed. Tetrodotoxin is active even after boiling for 1 hour, but it can be inactivated in acid and alkalis. Its mechanism of action is similar to saxitoxin. In humans, the symptoms — numbness of the lips, tongue, fingers and arms — begin a few minutes after ingestion of the toxin. Muscular paralysis and ataxia are followed by respiratory paralysis leading to death. Death may occur 30–60 minutes after consumption of 1–2 mg of tetrodotoxin.

Scombroid poisoning is an intoxication resulting from the consumption of inadequately preserved tuna, mackerel, shipjack and bonito, in which histamine and saurine are produced as a result of bacterial action. Scombroid fish apparently have a sharp or peppery taste. Signs and symptoms of intoxication include nausea, vomiting, diarrhea, epigastric distress, flushing of the face, headache and burning of the throat followed by numbness and urticaria (hives). Severe cases may develop cyanosis and respiratory distress and, although rarely, death may occur. The disease quickly responds to antihistamine treatment.

Shellfish poisoning is a paralytic disease resulting from the consumption of shellfish that have ingested toxic marine algae, especially the dinoflagellates *Gonyaulax catenella* and *G. tamarensis*. The toxins found in paralytic shellfish poison are saxitoxin, gonyautoxin II and III and neosaxitoxin. Saxitoxin is a neurotoxin that blocks the sodium pores of nerve and muscle membranes and prevents the passage of these ions, which are necessary for nerve conduction into the cells. Saxitoxin may cause death within a few hours after consumption of poisonous shellfish. Man is more susceptible to saxitoxin than mice are. Although boiling at neutral pH does not destroy all toxin, inactivation is enhanced by acid and alkaline pH. Boiling in bicarbonate-treated water and discarding the broth is suggested as a means of preventing shellfish poisoning. Consumption of 1 mg of the toxin can be mildly toxic, whereas 4 mg can be fatal if not treated vigorously. Toxic symptoms begin as a tingling sensation and numbness in the lips, tongue and fingertips; within a few minutes of eating toxic shellfish, numbness extends to the legs, arms and neck and is followed by general muscular lack of coordination that progresses to respiratory paralysis. Death may occur within 2–12 hours from respiratory failure, depending on the dose of the toxin. The recommended treatment for victims is artificial respiration. If death does not occur within 24 hours, normal functions can be regained within a few days.

OXALATES AND PHYTATES

Oxalates bind calcium and other needed trace minerals and make them unavailable for absorption. Increased oxalate intake can be responsible for hypocalcemia, which results in decreased bone growth. Deposition of insoluble calcium oxalate crystals in the kidneys may result in kidney stones. High levels can cause diarrhea, blood-clotting problems, convulsions and coma. Oxalates are naturally present in peas, cocoa, tea, spinach, berries, rhubarb, carrots, lettuce, turnips and beets. The high oxalate content of tea accounts for half of the dietary oxalate in areas where tea drinking is customary. If tea is made with hard water or served with milk, the calcium can precipitate oxalate from it. Rhubarb leaves contain toxic levels of oxalate, while the level of oxalate in spinach is up to 1% of fresh weight. Phytates bind divalent metals like zinc, cobalt, iron, calcium and copper, thus making them unavailable for absorption. They are primarily found in nuts, legumes and the germ and bran of cereal grains, but are also present in green beans, carrots, broccoli, potatoes, artichokes, blackberries, strawberries and figs. Mineral deficiencies may be observed in populations where cereals make up an important part of the diet.

PROTEASE INHIBITORS

Protease inhibitors are proteins that are unusually stable to proteolytic attack and some of them (e.g., soybean inhibitors) have shown growth-inhibitory and pathogenic properties in animals. The protein-digesting enzymes such as trypsin, chymotrypsin and others are examples of this group of natural toxicants. All plants, especially legumes, contain protease inhibitors. Common foods like beans (soy, mung, kidney, navy, lima), barley, beets, buckwheat, corn, lettuce, oats, peas, chickpeas, peanuts, potatoes, rice, rye, sweet potatoes, turnips and wheat also have protease inhibitors. Their major toxic effects can be summarized as impaired growth and food utilization, and pancreatic hypertrophy. In most cases, but not all, cooking destroys the inhibitory effects of these compounds. In the case of potatoes, microwave cooking and boiling are more effective than baking. Because many vegetables are eaten raw or after very short cooking times, their enzyme-inhibiting factors may not be inactivated.

SAPONINS

Saponins have a soaplike quality, from which they get their name. They are found in soybeans, sugar beets, peanuts, spinach, asparagus, broccoli, potatoes, apples, eggplant, alfalfa and ginseng root. They are capable of disrupting red blood cells and producing diarrhea and vomiting. They are highly toxic to cold-blooded animals; their toxic effects are related to the reduction of surface tension. Small concentrations are generally harmless to warm-blooded animals because intestinal microflora destroy them and blood plasma inhibits their action. The saponins in alfalfa leaves can be reduced by coagulation and washing at alkaline pH.

TANNINS

Tannins are found in almost every plant, but mainly in tea, coffee and chocolate. Bananas, sorghum, spinach, red wine and beer also contain important amounts of tannins. Tannins bind protein, precipitate the protein of the epithelium, penetrate through the superficial cells and cause liver damage. The 3–5% solution of tannin, toxic orally, retards growth. Tannins inhibit virtually every digestive enzyme and reduce the bioavailability of iron and vitamin B_1. Toxicity increases when they enter the blood stream. Tannins have been responsible for fatal poisoning of domestic animals fed acorns and other high tannin feeds like carob, certain sorghums, rapeseed meal and grapeseed meal.

TRISACCHARIDES

Raffinose and stachyose are nondigestible trisaccharides found mainly in legumes. The human body does not have the necessary enzymes to split these sugars, so they move unabsorbed into the large bowel and are fermented by the bacteria there, resulting in gas and abdominal rumblings. In some persons, they cause dizziness, headache and cramps.

VASOACTIVE AMINES

Vasoactive amines are phenylalkylamines that are present in foods such as aged cheeses, wine, beer, yeast extracts such as marmite, broad beans, potatoes, bananas, avocados, pineapples and pickled and fermented meats or fish. They are also found in large amounts of coffee and chocolate. These amines are related to the common aromatic amino acids; tyrosine, histidine, tryptophan, histamine, tryptamine, tyramine and serotonin can be given as examples. They cause facial flushing, rapid heart rate, increased blood pressure, headache and urticaria. Eating foods with tyramine, especially cheese and yeast products, while taking monoamine oxidase inhibitors (MAO) can lead to severe complications such as migraine attacks, intracranial bleeding and even death.

VITAMIN ANTAGONISTS

Vitamin antagonists are substances that cause toxicity by inhibiting the activity of vitamins. Citral from orange peel, orange and lemon oils, lipoxidase enzyme in raw soybeans and vitamin E antagonize vitamin A activity and cause changes in the eye similar to vitamin A deficiency. Avidin from raw egg, which binds biotin, and antipyridoxin factor linatine in flaxseed can also be given as examples of vitamin antagonists.

7

CONTAMINANTS

The term contaminant refers to undesirable materials that are present in food before, during or after processing. Joint FAO/WHO Codex Alimentarius Commission (CAC) defines contaminant as "any substance not intentionally added to food, which is present in such food as a result of the production, manufacture, processing, preparation, treatment, packing, packaging, transport or holding of such food, or as a result of environmental contamination." Different types of substances can contaminate our foods through the use of agricultural chemicals, by environmental chemicals, by the equipment and containers used during production and packaging, as well as by microorganism attack. The natural inorganic constituents in soil and water may also contaminate food products. Many toxic chemicals present in soil and water are absorbed by plants and by land and marine animals themselves. For example, selenium and nitrates may accumulate in plants to toxic levels. Indirect contamination of meats, milk or eggs may occur as result of ingestion of contaminated foods by animals. Microorganisms in the soil, water and air may infect the growing plant and stored products, and produce toxic metabolites. Contamination may also occur through the use of fertilizers, pesticides, growth stimulants and antibiotics used during the production of foods or from packaging or canning materials in contact with foodstuffs. Another source of chemical contaminants is the vessels and utensils used in cooking, food preparation and short-term household storage of prepared foods and beverages.

Contaminants are unintentional additives and their presence in foodstuffs should be guarded against, or else be kept at minimum levels. The best way to avoid the presence of contaminants in food is to act at the source to limit their presence in the environment. The European Community (EC) employs a wide variety of measures aimed at eliminating or limiting the presence of dangerous substances in the environment that might enter the food chain as contaminants. An example of the effectiveness of this policy is the decrease of the presence of lead as a contaminant in food following the progressive introduction of lead-free fuel for automobiles and trucks. Raw materials with levels of contaminants that could affect the safety of foods from a public health point of view should not be used for the manufacture of foodstuffs. Today, the introduction of contaminants into the food chain can be avoided by monitoring prevention systems such as HACCP (hazard analysis critical control points), beginning when the raw materials are purchased and at critical stages of the process until the foods reach the consumer.

The aim of good production practices and food regulations is to prevent the inclusion of contaminants, or at least to keep their amounts in foods below the toxic level. The safety assessment and limits of use of contaminants are evaluated by the Codex Alimentarius Commission (CAC) and the guideline levels adopted for different contaminants are intended for use in regulating foods moving in international trade. In 1972, the Joint Expert Committee on Food Additives (JECFA) established the concept of provisional tolerable weekly intake (PTWI), indicating the maximum acceptable weekly level of a contaminant in the diet, which is applied for contaminants with cumulative properties such as mercury, cadmium and lead. In this recommendation, tolerable intakes are expressed on a weekly basis because the contaminants given this designation may accumulate within the body over a period of time. Provisional tolerable daily intake (PTDI) is established for food contaminants that are not known to accumulate in the body, such as tin, arsenic and styrene. The value assigned to the PTDI represents human exposure as a result of the natural occurrence of the substance in food and in drinking water. Maximum tolerated dose (MTD) is defined as the "highest level of a food substance that can be fed to an animal without inducing obvious signs of toxicity other than those due to cancer." MTD has been set as the maximum dose level at which a substance induces a decrement in weight gain of no greater than 10% in subchronic toxicity tests. Countries all over the world can adopt these limits on a national level and establish their own regulations. The Scientific Committee for Food of the European Community is also conducting toxicological research and publishing reports related to different contaminants that are found in foods.

ANTIBIOTICS

Antibiotics are used in animal feeds to increase growth rate and improve feed utilization, as well as to protect animals from certain diseases. Penicillin, streptomycin and tetracycline can be given as examples of the antibiotics used for treatment of animals as well as in animal feeds. The use of antibiotics in therapy, prophylaxis and growth promotion of animals can result in residues in food, especially in milk. Generally, it is accepted that milk will contain residual antibiotics up to 72–96 hours after treatment of the animal with antibiotics. The levels of antibiotics in food may cause allergies and development of resistant organisms in humans. For example, highly sensitized persons may show allergic reactions to foods containing penicillin. Also, the overuse of antibiotics may contribute to the development of antibiotic-resistant bacteria. In an effort to minimize human exposure to antibiotics from feed, animal producers are required to follow prescribed withdrawal periods. These are based on tissue-clearance times and are set so that antibiotic levels drop below detectable levels. No measurable residues remain in milk or meat if the required drug withdrawal schedule is followed.

BACTERIAL TOXINS

Foods can become contaminated by bacteria in many ways, thus causing illness. Major factors that cause bacterial contamination during food production are inadequate cooking or reheating, or improper cooling of cooked foods. Cross-contamination between cooked and raw foods, inadequate cleaning of equipment and improper hot storage conditions are other important causes for bacterial contamination of foods. As a result, it can be stated that poor hygienic standards during food preparation and lack of training in food safety are the most important causes of food-borne illness. Bacterial food-borne disease may result from the presence in food of bacteria that can cause disease either by multiplying in the intestinal mucus or by toxins (enterotoxins) produced by microbes in the intestinal tract following multiplication, sporulation or lysis. Bacterial toxins are proteins that have an acute effect that occurs a few hours to a few days after exposure. Bacterial toxins that cause food-borne illness and the estimated number of cases of illness per year in the U.S. are listed in Table 7.1.

Table 7.1 Bacterial Toxins That Cause Food-Borne Illness

Toxin	Estimated Cases of Illness per Year in the U.S.
Staphylococcal enterotoxins	1,155,000
Bacillus cereus enterotoxins	84,000
Botulinal neurotoxins	270
Estimated total numbers of cases of food-borne illness in the U.S. from all causes	12,581,270

From Pariza, 1996, with permission of International Life Sciences Institute.

***Staphylococcus aureus* intoxication:** This type of toxicant is one of the most important causes of food-borne disease worldwide. *Staphylococcus aureus* is a gram-positive, nonmotile and nonspore-forming cocci that forms five enterotoxins (A-SEA, SEB, SEC, SED, SEE) that have different antigenic properties. Among these toxins, A-SEA is the most effective in causing food intoxications. *Staphylococcus* enterotoxins are simple proteins (250000–350000 MW) that have high heat stability. Generally, the enterotoxin production increases at optimum pH and temperature conditions. Foods containing *S. aureus* at a level of 5×10^5/g may possess a risk for *Staphylococcus* intoxication. However, even foods containing *S. aureus* at a very low level may involve the toxins because the heating operations applied to eliminate the bacteria may not affect toxins that are already formed. Poultry, meat, potato salad, cream-filled bakery goods and high-protein leftover foods, as well as sandwiches, puddings, ready-to-eat vegetable salads and pastries that are left at room temperature before serving are frequently involved in such intoxication. Other spoilage organisms present inhibit multiplication of *S. aureus* in raw food products. As a result, only cooked products subsequently contaminated by infected handlers and stored at warm temperature for several hours before consumption are capable of causing intoxication.

Signs and symptoms of *S. aureus* poisoning begin 1–6 hours after consumption of contaminated food. The major symptoms are nausea, vomiting, abdominal cramps and diarrhea. Sweating, dehydration, weakness, retching, salivation, anorexia and shock may also occur. Recovery usually occurs in 1–3 days after treating the patient by adjusting the liquid balance in the body. Fatalities are rare. Major preventive measures in reducing the incidence of *S. aureus* food intoxication involve education of food handlers regarding hygienic practices to reduce postcooking contamination of high protein foods and eliminating prolonged storage of cooked foods at room temperature before consumption.

***Clostridium botulinum* intoxication:** *Clostridium botulinum* is a gram-positive rod-shaped anaerobic bacteria capable of forming heat-resistant spores. The organism is widely distributed in soil, on raw fruits and vegetables and in the intestinal tracts of animals. Six types of heat-labile neurotoxins, A, B, C, D, E, F and G, have been isolated from various food products. Botulism is a neurotoxic syndrome caused by consumption of foods containing the neurotoxins produced by *C. botulinum*. Neurotoxins A and B are the types that are more likely to cause botulism in humans. Type E toxin is mostly isolated from the soil, but it can also be isolated from seawater, fish intestines and sea or lake sediments. Botulism has been linked to coleslaw prepared from packaged shredded cabbage. Studies have shown that *C. botulinum* can produce toxin in polyvinyl film-packaged and vacuum-packaged mushrooms. It is important that the permeability characteristics of packaging films minimize the possibility of development of anaerobic conditions suitable for growth of clostridial spores. Because anaerobic pockets can develop in tightly packed produce, even when films have high rates of oxygen and carbon dioxide permeability, an additional measure to prevent growth of *C. botulinum* is to store products at less than 3°C. A study conducted in the U.S. has shown that 72% of botulism incidences evolved from home-prepared canned products, and most outbreaks have been caused by B and E toxins. Moisture, a pH above 4. 6 and storage under anaerobic conditions for a period of time are required for *C. botulinum*-contaminated foods to accumulate sufficient quantity of the toxin. If such a product is consumed without sufficient reheating, the likelihood of botulism increases. No food would cause a botulism problem if all *C. botulinum* spores were destroyed before they could germinate into a vegetative culture. The most practical way of achieving this is heating the food sealed in a container. In a canning operation of low-acid foods, minimum heating time and temperature is (12D) 2.4 minutes at 121°C (F0 = 2.4), which will inactivate 10^{12} botulinum spores to 1.90% of the spores present. Botulinum toxin is a super toxic compound having an LD_{50} value of 0.00001 mg/kg body weight. Heating at 79°C for 20 minutes or at 85°C for 5 minutes destroys preformed toxins. Another factor in the prevention of botulinum toxin formation in cured meats is the concentration of nitrite. An inverse relationship exists between the concentration of nitrite initially added to the food and the probability of toxin formation.

Common foods involved in botulism are home-canned vegetables such as beans, corn, leafy vegetables and, especially, peppers, all of which contain toxins A and B. Canned fruits are also involved, though to a lesser extent. Type E is isolated mostly from fish products and type F from liver paste. All of these toxins can cause botulism in humans. Signs and symptoms of botulism usually appear 12–24 hours following consumption of contaminated food. Initial signs of nausea, vomiting and diarrhea are followed later by predominantly neurological signs including headache, dizziness, blurred vision, weakness of facial muscles, loss of light reflex and pharyngeal paralysis. Paralysis of the respiratory muscles leads to failure of respiration and death, usually in 3–5 days. Death may also occur in periods from 24 hours to 2–3 weeks. In survivors, partial paralysis might persist for 6–8 months and recovery is associated with growth of new fibrils outside the area of old fibrils. Food-borne botulism can be prevented by proper home preservation of foods, as well as commercial canning, boiling vegetables for at least 3 minutes before serving, discarding all swollen and damaged canned products after boiling and the use of a *C. botulinum* toxoid to protect individuals involved in research and testing. In smoked-fish products, application of the smoking operation to heat the middle region of the fish to 82°C for at least 30 minutes and immediate freezing of the product are essential for preventing botulism. Adjustment of pH, salting, or use of chemical preservatives such as sodium or potassium nitrates and potassium nitrites are other treatments that can be performed to inhibit the formation of *C. botulinum.*

***Bacillus cereus* intoxication**: *Bacillus cereus* is a gram-positive, rod-shaped, spore-forming aerobe widely distributed in nature and most commonly found in soil, milk, cereals, starches, herbs, spices and other dried foods. Some strain of *B. cereus* can grow at refrigeration temperatures. Foods other than raw fruits and vegetables are generally linked to illness implicating *B. cereus*. Illness associated with eating contaminated mustard and cress sprouts has been reported. *B. cereus* synthesizes two types of enterotoxins (emetic and diarrheagenic enteroxins), therefore, two types of intoxication are caused. Diarrheagenic toxin is a protein that is produced by actively growing cells. It is activated by enzymes such as trypsin and pronase, as well as by exposure to 56°C or more for 30 minutes. Emetic toxin is stable to heat, pH, and tripsin and pepsin enzymes.

The majority of food-borne disease outbreaks caused by *B. cereus* intoxications have occurred in northern and eastern Europe. A diarrheal illness involving a wide variety of meats and vegetables, various desserts, fish, pasta, milk and ice cream (similar to that of *C. perfringens*) and a vomiting illness involving cereals and fried rice served in Chinese restaurants are both apparently caused by *B. cereus.*

***Clostridium perfringens* intoxication**: *Clostridium perfringens* frequently causes food-borne infections that subsequently lead to sporulation of the organism in the large intestine. It also causes gangrene, appendicitis, puerperal fever and enteritis in humans by nonfood-borne routes. The enterotoxin (lecithinase), released during sporulation of the bacteria, is capable of causing fluid accumulation in the intestines. Lecithinase, also called α-toxin, possesses lethal-necrotizing as well as hemolytic activities and is produced in greatest quantities by type A organisms. Among the five antigenically (toxicologically) distinct types of *C. perfringens* (type A through E), type A is almost always involved in food-borne gastroenteritis in humans in the United States. Type C produces two different (lethal-necrotizing and hemolytic) toxins and has caused only two outbreaks of sometimes fatal enteritis necroticans in Europe. Only meat and fish products are involved, due to the availability of all the amino acids and growth factors required for growth of *C. perfringens*. Roast beef, beef stew, gravy and meat pies for type A and pork, other meats and fish for type C are frequently involved. Typically, foods involved are cooked at <100°C for less than an hour and are subsequently kept warm or slowly cooled. Spores that survive the heat shock multiply faster in the food than those not subjected to heat treatment and also are capable of elaborating greater quantities of entero-toxin in the gut.

HORMONES

Hormones, regulators of physiological processes by chemical means, are widely used as growth promoters in animals in agriculture today. Because some proteins, peptides and steroidal hormones act as anabolic agents, they have been used for this purpose. The protein and peptide hormones have little significance to man because they are denatured and biologically inactivated during the digestive process. In contrast, steroidal hormones are readily absorbed from the gastrointestinal tract, maintaining their effectiveness until they are metabolized to inactive compounds. During the past few decades, many other chemicals have been shown to possess hormone activity and some have also been used with the aim of controlling reproductive function in breeding animals and as anabolic agents. Natural hormones, such as progesterone, are used to regulate pregnancy in animals. Because the residues of these hormones in edible portions of the animals are at very low levels, the use of these natural hormones or their derivatives are not important from a toxicological point of view.

Stilbens, a group of synthetic hormones that show estrogenic activity, are used for increasing weight and feed gain as well as for treatment purposes in zootechnical studies. A typical hormone in this group is diethylstilbestrol (DES), which is a synthetic estrogen-like substance that was used during and after the 1950s to promote enhanced tissue accumulation and faster weight gain in cattle and sheep. However, when this hormone was given to pregnant women to prevent miscarriage, vaginal cancers occurred in some of the young adult daughters of these women. So, in the U.S., a strict protocol is required for the removal of DES before animals are sent to market. There are also nonstilben types of synthetic hormones such as melangestrol acetate (MGA) and zeranol. Melangestrol acetate, first synthesized in 1960, is used as a growth promoter. Zeranol (7α-zearalanol) is an estrogenic hormone that is obtained from the mycotoxin zearalenone. It is also used for increasing weight gain in animals, and has a toxic effect on the liver at all doses, while causing hematological changes at high doses.

MYCOTOXINS

Mycotoxins are highly toxic compounds of small molecular weight produced by molds or fungi. Many of the molds capable of producing mycotoxins are contaminants of food and agriculture commodities. They are capable of growing in a variety of places and under several moisture and temperature conditions. Thus, most foods are susceptible to invasion by molds during some stage of production, processing, transport or storage. If mold growth occurs, mycotoxin can possibly form and toxins may persist long after the actual molds have disappeared from the foods. The physical parameters that affect mycotoxin formation are fungal growth, moisture content and relative humidity. On the basis of moisture requirements and time of appearance in the production cycle of grain, fungi can be divided into two groups — field fungi and storage fungi — with some organisms overlapping into both. Field fungi that have high moisture requirements invade seeds prior to harvest and are best represented by *Fusaria, Alternaria* and some *Penicillia*. Field fungi require a moisture content of 22–23% and a relative humidity of 90–100%. Storage fungi are primarily *Aspergilli* and some *Penicillia*. For growth, these organisms require grains whose moisture content is generally above 15% and in equilibrium with a relative humidity of 70–90%. The optimal water activity value for toxin production by toxigenic fungi is in the range of 0.93–0.98. The limiting a_w value, which is the value below which mycotoxins are not produced, is in the range of 0.71–0.94, depending upon the mold in question. The general pH value range of the substrate favorable for toxin production is 3.4–5.5, although some toxigenic molds can initiate growth at the low pH values, causing the pH value of the substrate to increase and become favorable for mycotoxin production. The most commonly found mycotoxins in foodstuffs are aflatoxins, citrinin, ergot, fumonisins, ochratoxins, patulin, sterigmatocycstin, trichothecenes and zearalenone. Under laboratory conditions, hundreds of other mycotoxins have been also produced and described.

Aflatoxins

Aflatoxins are the group of mycotoxins that has attracted the most attention since they were first discovered in 1960. They are unavoidable contaminants of various foods and incidences of aflatoxin contamination are common in temperate countries, whereas they are usually less frequent in cooler countries, particularly before harvest. Aflatoxins are produced by some strains of *Aspergillus flavus* and *Aspergillus parasiticus,* which can grow on certain foods at favorable conditions of temperature and humidity. There are mainly six aflatoxins of concern: B_1, B_2, G_1, G_2, M_1 and M_2. The M toxins were first isolated from milk of lactating animals fed with aflatoxin B_1-contaminated feed. Aflatoxin B_1, the principal member of the aflatoxins, has been responsible for numerous deaths of poultry and other domestic animals. A potent liver toxin, its toxicity tests have shown that it causes liver cancer in experimental animals. Aflatoxin M_1, a metabolite of aflatoxin B_1, is found in the milk of dairy cattle that have ingested moldy feed. If a mammal such as a dairy cow consumes aflatoxin B_1 in the diet, about 1–3% of the amount of B_1 that is consumed will appear in milk as aflatoxin M_1. Aflatoxin M_1 is stable in raw milk and processed milk products and is unaffected by pasteurization or processing into cheese or yogurt. The ability of this toxin to induce tumors in experimental animals and the relatively large consumption of milk by children has made it a food contaminant of worldwide concern. On the other hand, eating meat from animals that have consumed aflatoxin-contaminated feed is of lesser significance. The ratio of aflatoxin B_1 in the dairy to aflatoxin M_1 found in milk is found to be approximately 75:1. The ratio of aflatoxin concentration in feed to that in eggs and the livers of certain animals was determined to be higher.

Because aflatoxins are considered potent mutagens, teratogens and carcinogens, several studies have been made to determine inhibition of their formation as well as the degradation of aflatoxins that are already present in foods. Thus, degradation of aflatoxins by other treatments has also been studied. Among these treatment methods, the use of ultraviolet radiation, heat, oxidizing agents such as hydrogen peroxide, sodium hypochloride, or exposure to alkaline substances like ammonia, sodium bisulfite and sulfur dioxide gas can be given as examples. Among these agents, the use of ammonia was commercially applied in detoxifying cottonseed and corn in the U.S. However, toxic residues and objectionable changes in the sensory and nutritional quality of decontaminated materials have occurred. The effects of aflatoxin degradation techniques are still investigated and there is need for new toxicological data on the substances that are formed as a result of the degradation treatments.

Citrinin

Citrinin is a yellow compound produced by several *Penicillium* and *Aspergillus* species. *Penicillium citrinum* is widely distributed in all rice-producing areas of the world. It has been isolated from rice produced in China, Burma, Egypt, Italy, Japan and the U.S. The mold commonly grows on stored rice, particularly on polished rice. Growth is commonly accompanied by formation of the pigment that causes the surface of rice kernels to appear yellow. An acute form of cardiac beriberi known as "yellow rice syndrome" was first implicated in Japan as early as the 17th century, when isolates of *P. citrinum* produced high quantities of this yellow mycotoxin. Citrinin is a strong kidney toxin that has also been identified as a contaminant of yellow peanut kernels from damaged pods. The toxicity of citrinin is low compared with ochratoxin A. Citreoviridin is another yellow rice toxin produced by *Penicillium citreoviride*. In animals, the major symptoms that begin with the ingestion of this toxin are paralysis of the hind legs and flank, vomiting and convulsions. These symptoms are followed by gradually developing respiratory disorders, cardiovascular disturbances, paralysis, decrease in body temperature, gasping, coma and respiratory arrest.

Ergot

The history of human mycotoxicosis dates back to the Middle Ages, where epidemics of ergotism (St. Anthony's fire) have occurred in Central Europe since the 1200s and earlier. Ergotism, which is now rare, was first associated with the consumption of scabrous (ergotized) grain in the mid-16th century. The fungal agents responsible for ergotism are *Claviceps purpurea* and *Claviceps paspali*, which infect mainly rye, oats, wheat and barley. Lysergic acid derivatives, the amine and the amino acid alkaloids of ergot, are identified as the causative agents of the gangrenous (*C. purpurea*) and nervous (*C. paspali*) forms of the disease. A typical symptom of gangrenous ergotism is a burning sensation in the feet and hands, followed by restriction of blood to these extremities, which results in gangrene. The symptoms of convulsive ergotism begin with hallucinations and involve neurological disorders, numbness, cramps, severe convulsions and death. Both syndromes have been documented in the recent literature in domestic animals consuming ergotized grains and in humans treated with ergotamine for migraine headaches.

Ergot alkaloids are smooth-muscle stimulants, promoting vasoconstriction (leading to gangrenous ergotism) and inducing uterine contractions (oxytocic effect). Ergot alkaloids can antagonize serotonin and block both the stimulatory and inhibitory CNS responses of epinephrine. The U.S. Department of Agriculture (USDA) grains division has set a tolerance limit of 0.3% (by weight) of contaminated grain in commercial trade.

Fumonisins

The fumonisins are a group of recently characterized mycotoxins produced by a limited number of *Fusarium* molds, of which *F. moniliforme* and *F. proliferatum* are the most important, as they frequently infect corn crops around the world. The contamination of foods and feeds usually reflects the degree of fungal infection of the original crops during a particular season, the incidence of which is influenced by various factors such as origin, drought stress and insect damage. At least three naturally occurring fumonisins occur, known as B_1, B_2 and B_3. Fumonisin B_1 is always the most abundant (representing approximately 70% of the total concentration) in naturally contaminated foods and feeds. These mycotoxins produce a wide range of biological effects, including leukoencephalomalacia in horses, pulmonary edema in pigs and nephrotoxicity and liver cancer in rats. Although their effects on humans are difficult to determine, fumonisins have been statistically associated with a high incidence of esophageal cancer in certain areas of Transkei, South Africa and could also play a role (along with the trichothecene deoxynivalenol) in the promotion of liver cancer in certain areas of China where they are endemic. On the basis of available toxicological evidence, the International Agency for Research on Cancer (IARC) has declared *F. moniliforme* toxins to be potentially carcinogenic to humans.

Ochratoxins

Aspergillus ochraeus and other Aspergillus species, as well as *Penicillium viridicatum* and *P. cyclopium* produce these toxins. They can contaminate corn, pork, barley, wheat, oats, peanuts, green coffee and beans. The main toxin of this group, ochratoxin A, can cause kidney damage in rats, dogs and swine. It is thought to be involved in a disease of swine known as porcine nephropathy, which is linked to moldy barley. Signs include lassitude, fatigue, anorexia, abdominal pain and severe anemia followed by signs of renal damage. It was also reported that ochratoxin A has association with chronic nephropathy (Balkan or endemic nephropathy), which was seen in Yugoslavia, Romania and Bulgaria in the years 1957 and 1958. Ochratoxin A in animal feed can be transmitted to meat, body fluids and eggs. The studies have shown that 2–7% of ochratoxin A is transmitted to beer during processing; more than 80% of ochratoxin A is destroyed on roasting of coffee; variable losses of ochratoxin A occur when baking with toxin-contaminated flour.

Patulin

Patulin is toxic to many biological systems, including bacteria, mammalian cell cultures, higher plants and animals, but its role in causing animal or human disease is unclear. It is produced by various *Penicillium* and *Aspergillus* species. *P. expansum* is the principal cause of apple rot and a common pathogen on many fruits and vegetables. It is found in moldy apples, plums, peaches, pears, apricots, cherries and grapes. Mold growth and subsequent production of patulin normally occurs only where the surface tissue of fruit has been damaged, although the presence of patulin in otherwise visibly healthy fruit can be discounted. This surface damage may be caused by insect or storm damage and handling procedures. Patulin is also a contaminant of fruit juices, especially apple juice. It is unstable in the presence of sulfydryl compounds and sulfur dioxide;

pasteurization does not destroy this mycotoxin but more than 99% of patulin is destroyed during fermentation. The effect of fermentation in destroying patulin was detected during production of apple cider from contaminated apple juice. Studies have shown that patulin has mutagenic, carcinogenic and teratogenic effects.

Sterigmatocystin

Fungi such as *Aspergillus versicolor, A. sydowi, A. nidulans, Bipolaris* spp., *Chaetomiun* spp. and *Emericella* spp. produce sterigmatocystin. This toxin was first detected in brown rice stored in warehouses under natural conditions. It is also found in moldy wheat and green coffee beans. Structurally related to aflatoxin, this toxin is also a liver carcinogen and also shows mutagenic effects. Some reports link sterigmatocystin and gastric cancer in China, although low vitamin C and other dietary factors have also been implicated.

Trichothecenes

The trichothecenes are a group of closely related metabolites produced essentially by a broad range of *Fusarium* molds, of which *F. graminearum* and *F. culmorum* are the most important. These fungi typically develop during prolonged cool, wet growing and harvest seasons to produce fusarium head blight (known as scab in the U.S.) in cereal crops. More than 20 naturally occurring trichotocenes are produced by *Fusarium* species, including T-2 toxin, neosolaniol, diacetylnivalenol, deoxynivalenol (DON or vomitoxin), HT-2 toxin and fusarenon. T-2 toxin, which is produced primarily by *F. sporotrichioides* and *F. poae,* is quite toxic to rats, trout and calves. It is also thought to be involved in the human disease known as alimentary toxic aleukia (ATA), which affected thousands of people who ingested over-wintered grain in parts of the Soviet Union during World War II. The disease was characterized by total atrophy of the bone marrow, agranulocytosis, necrotic angina, sepsis, hemorrhagic diathesis and mortality ranging from 2–80%. It was linked to the consumption of over-wintered cereal grains and wheat or bread made from them. Although T-2 toxin is rarely found in foods, there is significant contamination in animal feeds.

F. nivale produces a group of trichothecene toxins including nivalenol and fusarenon-X in the flowering grainhead of wheat, barley, rice, corn, other cereals and certain forage grasses. The disease in cereal grains called red-mold disease (*akakabi-byo*) or black spot disease (*kokuten-byo*) has been associated with intoxications in humans, horses and sheep in Japan. Symptoms in humans include headaches, vomiting and diarrhea, with no fatalities. *F. roseum,* capable of producing deoxynevalenol and its acetylated derivatives on rice and barley, was also isolated — suggesting multiple causes.

Zearalenone

This toxin is produced from *Fusarium* species and occurs naturally in high-moisture corn in late fall and winter, primarily from the growth of *F. graminerium (roseum) and F. culmorum.* Zearalenone, also known as F-2 toxin, is an estrogenic compound that causes vulvovagitinis and estrogenic responses in swine. It produces metabolites that act like the hormone estrogen and cause hyperestrogenic syndrome. High levels of these metabolites cause infertility in both males and females. Zearalenone contamination primarily occurs in corn, but this toxin may also be present in beer, cassava, beans, walnuts, bananas and soybeans. Production of the toxin on corn or other cereals is favored by temperatures near freezing for an extended time and also by temperature cycling from low to moderate temperature and back again. The mold can also grow on corn during storage if it is harvested with too much moisture and is not dried properly before storage.

Miscellaneous Mycotoxins

The properties of mycotoxins, other than those described above, that have different types of toxicological significance are shown in Table 7.2.

N-NITROSO DERIVATIVES

N-Nitroso amines (nitrosamines) after metabolic activation, and N-nitroso amides (nitrosamides) are strong alkylating agents. Many of these compounds are potent carcinogens. Nitrosamides are direct-acting carcinogens, while nitrosamines are secondary carcinogens that require enzymatic activation for their carcinogenic action. They produce cancers in a variety of organs of the body including the esophagus, stomach, liver, lower gastrointestinal tract, pancreas and in the respiratory tract. Research conducted on this subject has shown that nitrites, commonly added to meats during the curing process, as well as nitrites formed by the bacterial reduction of nitrate in the mouth, react, under appropriate conditions, with amines or amides that are present in food as degradation products of proteins or other food components to yield nitrosamines or nitrosamides. The consumption of nitrates and nitrites in food and water, the facile reduction of nitrates to free amines as well as those compounds that act as nitrosating agents, are widespread in nature and thus, N-nitrosamines have been discovered in a

Table 7.2   Properties of Selected Mycotoxins

Mycotoxins	Major Producing Fungi	Typical Substrate in Nature	Biological Effect
Alternaria mycotoxins	*Alternaria alternata*	cereal grains, tomato, animal feeds	mutagenic, hemorrhagic
Cyclopiazonic acid	*A. flavus, P. cyclopium*	peanuts, corn, cheese	neurotoxins, cardiovascular lesion
Cyclochlorotine	*P. islandicum*	rice	hepatoxin, carcinogenic
Luteoskyrin	*P. islandicum, P. rugulosum*	rice, sorghum	hepatoxin, carcinogenic, mutagenic
Moniliformin	*F. moniliforme*	corn	neurotoxin, cardiovascular lesion
Penicillic acid	*P. puberulum, A. ochraceus*	barley, corn	neurotoxin, mutagenic
Penitrem A	*P. palitans*	foodstuffs, corn	neurotoxin
Roquefortine	*P. roqueforti*	cheese	neurotoxin
Rubratoxin	*P. rubrum, P. purpurogenum*	corn, soybeans	hepatoxin, teratogenic

Adapted from Chu, 1995 and Pariza, 1996.

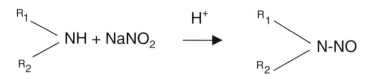

Figure 7.1   Nitrosamine formation (adapted from Anselme, 1979).

variety of foods and beverages and even baby bottle nipples. Figure 7.1 shows nitrosamine formation by nitrosation of secondary amines with sodium nitrite.

The presence of nitrosamines in food became apparent as a result of research that showed high incidence of liver disorders in mink and ruminants traced to the consumption of a herring meal that was treated with high levels of sodium nitrite. N-Nitrosodimethylamine, one of the most powerful carcinogens known to man, was isolated from the herring meal. It was postulated that the free amines present in the fishmeal had undergone nitrosation, resulting in the presence of nitrosamines. Nitrates and nitrites occur to some extent in almost all water sources, and these ions are readily taken up by growing plants. N-nitrosodimethylamine is the most frequently found volatile nitrosamine in cheeses, beer and nitrate-nitrite-preserved meats.

Some nitrate-nitrite meats also contain N-nitrosopyrrolidine and N-nitrosopiperidine, and the N-nitrosopyrrlodine content of meat frequently increases on cooking. Because many N-nitrosamines and amides are potent carcinogens, it is necessary to eliminate or decrease the levels of human exposure to these compounds. Contamination of beer with nitrosamines provided an important source of volatile nitrosamine intake in some countries like Germany, but modification of the malting procedure to avoid contact of the malt with hot air containing nitrogen oxides has largely eliminated this route of contamination.

Because the use of nitrites and nitrates cannot be banned unless alternatives are found to combat the threat of *C. botulinum* growth in their absence, sodium ascorbate or sodium erythorbate are being used in cured-meat products to inhibit nitrosamine formation.

PACKAGING MATERIALS

One of the primary functions of packaging is to protect food from contamination to preserve its safety and quality. With increased urbanization and growth of the food industry, more and more foodstuffs are prepackaged before reaching the retailer's shelves. Many problems are involved in packaging because almost any packaging material is subject to slow chemical or physical attack from the food or storage conditions. When some of the packaging material becomes a part of the food as consumed, it is termed a migrant and is classifed as an incidental or unintentional additive. It should be controlled by considering the related regulations.

The traditional metal containers that came into use in the 19th century have been improved over the years so that the contents are minimally exposed to lead solder. The one exception appears to be canned milk, which can be packaged in a can with a flat-soldered seam and the filling hole closed by solder flux. The lead content of canned evaporated milk has been found to be higher than that present in either fresh or powdered milk. Although no cases of lead poisoning in infants have been traced to the use of canned milk, there is concern that the lead content of canned milk could increase the danger to children at an age when they are particularly susceptible to lead poisoning.

The packaging of foodstuffs in plastic materials presents very complex problems that are being studied at the present time. Although the plastic itself may be relatively insoluble, partially reacted polymers, plasticizers and contaminants can be dissolved and migrate into the food or become environmental problems. Phthalic acid esters (PAEs), which are extensively used as plasticizers in foodwrap films, have been found in whole blood stored in plastic bags. PAEs are now widely dispersed throughout the environment, but their significance as environmental pollutants is not fully known. The use of polyvinyl chloride (PVC) plastic bottles for packaging of liquors was banned in the U.S. in 1973 because of possible toxic interaction between the alcohol and PVC. In recent years, much research has been conducted on the safety of heat susceptors used for microwave foods. Microwave heat susceptors are metalized surfaces that are components of food packages intended for microwaving to promote the browning and surface crispness of the food. There is concern that, at high temperatures (> 500°), the degradation products of these compounds might migrate into the foods. The high temperatures attained by packaging using susceptor technology may result in the formation of significant numbers of volatile chemicals from the susceptor components, as well as loss of barrier properties of food-contact materials, which could lead to rapid tranfer of nonvolatile adjuvants to foods. Studies by the FDA, with hot vegetable oil in contact with a susceptor, have shown that the susceptor materials liberate volatile chemicals that may be retained in the oil at parts-per-billion (ppb) levels.

Recently, concern about contamination of foods with bisphenol A (BADGE), a monomer in the form of diglycidyl ether that is used in the production of epoxy resins and polycarbonate plastics has gained importance. The molecular structure of bisphenol A is shown in Figure 7.2.

This monomer may leach out of the plastic due to incomplete polymerization, or because of breakdown of the polymer upon heating, for example, during sterilization by autoclaving. The plastics are utilized in food and drink packaging applications to line metal food cans, bottle tops and water supply pipes. This substance is also widely used in households and industry. It can be expected to be present in sewage and wastewater effluents and concentrated in sewage sludge. Studies have shown that bisphenol A has leached from cans into the vegetables they contained. Bisphenol A has been shown to possess estrogenic properties, and high doses of this substance were reported to cause reproductive toxicity and affect cellular development in rats and mice. Bisphenol A also showed cytotoxic and genotoxic effects *in vitro*.

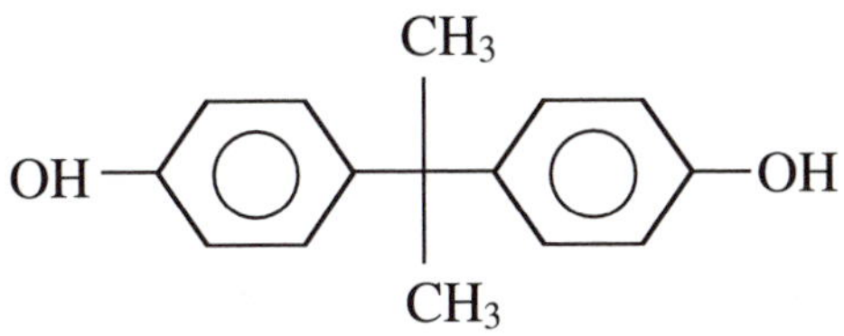

$$\text{OH}-\text{\textcircled{}}-\overset{\overset{\displaystyle CH_3}{|}}{\underset{\underset{\displaystyle CH_3}{|}}{}}-\text{\textcircled{}}-\text{OH}$$

Figure 7.2 Molecular structure of bisphenol A (adapted from Biles et al., 1997).

Table 7.3 Functions and Examples of Different Categories of Some Pesticides with Significant Toxicity

Category	Function	Examples
Halogenated hydrocarbon pesticides	Insecticides	DDT, dieldrin, aldrin, endrin, chlordane, heptachlor, kepone, lindane, mirex
Organophosphate pesticides	Insecticides	malathion, parathion
Carbamates	Insecticides, herbicides, fungicides	aldicarb, carbaryl, ethylene bisthiocarbamate
Alkyl phenolic acid herbicides	Herbicides	2,4-D, 2,4,5-T

Adapted from Smith, 1995.

PESTICIDES

Pesticides are chemicals or mixtures of chemicals used for the prevention, elimination or control of unwanted insects, plants and animals. CAC defines pesticides as "any substances or mixture of substances intended for preventing or controlling any pest" and includes any substance or mixture of substances intended for use as a plant growth regulator, defoliant or desiccant. Although pesticides have several benefits, the residues of some of them present important toxic problems in foods. Table 7.3 shows categories and functions of some pesticides with significant toxicity.

The organochlorine pesticides, the most persistent substances in the environment, tend to concentrate in the edible parts of plants and the fatty tissues of numerous animal species and humans. Many are banned in several countries. For example, the organochlorine insecticide DDT was once praised as a perfect pesticide because of its important effects against several pests, but, since its carcinogenic potential on rats and mice was discovered, there have been suspicions about its use. Other organochlorine types of insecticides — dieldrin, endrin and aldrin — like DDT form an enterohepatic cycle in the body and have even higher acute toxicity than DDT. Chlordane, heptachlor, lindane, mirex, TCDD and kepone were also found to be unsafe.

Organophosphate types of pesticides are more acutely toxic than organochlorines. The toxicity of the compounds in this group varies according to their structures and functions. The most toxic organophosphate type of pesticide is parathion. Malathion is toxic to insects but much less toxic to humans than parathion.

Carbamates are often used when insects have become resistant to organophosphates. Their major toxic symptoms are constriction of the pupils, muscle weakness, spasms, lowered blood pressure, respiratory failure, convulsions and cardiac arrest. The most toxic carbamate is aldicarb, a systemic pesticide that cannot be washed off, however, heating reduces the residue level.

Among the fungicides, captan is shown to be toxic to fish especially. Its toxicity increases when diets are low in protein. The fungicide ethylenebisdithiocarbamate has relative toxicity, but it has antithyroid activity. Alkyl phenolic acid herbicides, such as 2,4-D and 2,4,5-T show greater toxicity than DDT. Animal feeding studies have shown that they caused decrease in total weight gained, increased kidney weight and hypertrophy of liver cells in animals. The herbicide known as dioxin or 2,3,7,8-TCDD (2,3,7,8-tetrachlorodibenzo-p-dioxin or TCDD) is an important carcinogen, usually found as the toxic impurity of the herbicides 2,4-D and 2,4,5-T. Dioxin causes cancer at high levels to workers who are exposed to it during its application to plants. The tolerable daily intake (TDI) proposed for this substance is as 1–4 pg/kg bw.

The most toxic compound detected among the fumigants is ethylene dibromide, whose carcinogenic and mutagenic activities have been reported. Daminozide (alar) is a plant growth regulator that was shown to cause various types of cancers. It is a systemic pesticide, meaning that washing does not remove its residues from the plants. A group of pesticides previously approved by the Joint FAO/WHO/CAC were later withdrawn, but continue to persist in the environment as contaminants. These include aldrin, dieldrin, chlordane, DDT, endrin and heptachlor. For these pesticides, the Joint FAO/WHO Meeting on Pesticide Residues (JMPR) converted maximum recommended levels (MRLs) to extraneous residue limits (ERLs). ERLs are established on reported levels of pesticides in food commodities, often through the use of data provided by the Global Environment Monitoring System/Food Contamination Monitoring and Assessment Programme (GEMS/Food). As the pesticides degrade in the environment, ERLs may be lowered as warranted by contemporary monitoring data. It is believed that the establishment of ERLs helps to protect public health, avoids unnecessary trade restrictions and discourages the direct use of the compound on food crops.

POLYCHLORINATED BIPHENYLS (PCBs)

The polychlorinated biphenyls (PCBs) consist of eight types of compounds, depending on the number of chlorine substituents. There are theoretically approximately 210 possible chemical isomers that can be produced from the chlorination of biphenyls; these isomers range from the dichloro to the nonachloro derivatives. The molecular structure of 2,4,3´,4´-tetra-chlorobiphenyl is demonstrated in Figure 7.3.

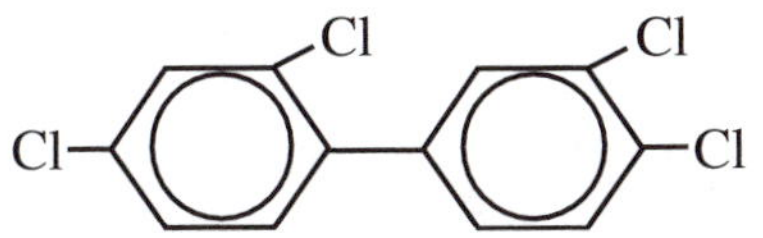

Figure 7.3 Molecular structure of 2,4,3',4'-tetra-chlorobiphenyl (from Concon, 1998b, with permission from Marcel Dekker, Inc.).

PCBs are used industrially as dielectric fluids in transformers, as plasticizers, as heat transfer and hydraulic fluids. They are characterized by their high chemical stability in water, acids, alkalis and high temperatures. The commercial product is marketed as a mixture of several chloroderivatives under different trade names: Arochlor (U.S.), Kanechlor (Japan), Phenochlor (France) and Colphen (Germany). The widespread use of these compounds by different industries causes environmental contamination through industrial leakages and wastes. They contaminate foods by way of water and soil. The first cases of human poisoning caused by PCBs were first recognized in Japan in 1968, where rice bran oil was contaminated with leakage from the heat exchanger of the equipment used to deodorize the oil of the commercial PCB mixture Kanechlor 400. The poisoning, which was called *yusho* (Japanese for "oil disease"), produced severe symptoms of acne (chloracne), weakness, numbness of limbs, swelling of eyelids, eye discharges, dark skin pigmentation and liver damage. PCBs can migrate from packaging material into food. An important contamination incident was reported in rice oil when recycled paper including PCB-containing carbonless paper was used to make cartons for packaging. Fish can accumulate significant amounts of PCB if the waters in which they live are contaminated. Cooking can reduce the PCB content of fish or other foods. Acute exposure to PCBs causes chloracne with dark brown pigmentation, itching, skin rash, burning of eyes, nose and throat and dizziness. If breast milk is contaminated, motor immaturity and adverse effects in visual memory can be caused in infants. PCBs can pass the placental barrier and can show teratogenic effects. Chronic exposure of experimental animals to PCBs has been shown to increase cancer risk. Studies also have shown that higher storage levels of vitamin A protect these animals from the adverse effects of PCBs.

POLYCHLORINATED DIBENZODIOXINS AND DIBENZOFURANS (PCDDFs)

Polychlorinated dibenzodioxins and dibenzofurans, a mixture of 210 different chlorinated congeners, are important pollutants of the environment, food and public health care. Environmental and toxicological hazards differ by several orders of magnitude, although they reflect the degree of chlorination. From the total abundance of 75 chlorinated dibenzodioxins and 153 chlorinated dibenzofurans, the 17 substituted 2,3,7,8-congeners (7 dioxins, 10 furans) with 4–8 chlorine atoms are subject to special significance due to their toxicity and environmental stability. PCDD/Fs are released in traces from nearly every combustion process involving chlorine and may be unintentionally synthesized in several processes in chlorine chemistry. The atmospheric deposition on plants and forage plants leads to the nonoccupational exposure of consumers and farm animals. Within the food chain, milk fat is exposed by accumulation from contaminated feed consumed by cows. The carryover rates to milk fat and other food of animal origin are congener-specific and are between 0 and up to 40%. The molecular structures of dibenzo-p-dioxins (PCDDs) and dibenzofurans (PCDFs) are shown in Figure 7.4.

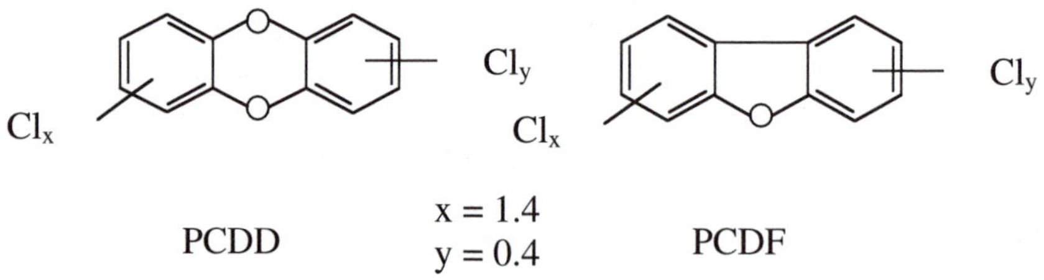

Cl_x PCDD Cl_y Cl_x

x = 1.4
y = 0.4

PCDF Cl_y

Figure 7.4 Molecular structures of polychlorinated dibenzo-p-dioxins (PCDDs) and dibenzofurans (PCDFs)(Anon, 1997b, with permission of International Dairy Federation).

The assessment of these substances as hazardous environmental chemicals of highest priority was emphasized by the Seveso accident in the summer of 1976. This occurred in the Italian city of Seveso when chemicals were released through an uncontrolled exothermic reaction from a chemical plant producing pesticides. The toxic clouds contaminated the surrounding areas and 200,000 people were affected by inhalation of the toxic chemicals. Among the chemicals, dioxin (2,3,7,8-TCDD, or tetrachlorobenzodioxin), has received the most attention, as it is considered to have the greatest toxic potency of all the chlorinated dioxins and furans. The major acute symptoms of toxication were skin changes, and the delayed acute effects after 20 years have been reported as carcinogenesis and teratogenesis. The 2,3,7,8-TDD congener is also the chemical that is equated with the generic name dioxin, or Seveso toxin. The toxic potential of chlorinated dioxins and furans varies with the different isomers. To compare various mixtures of dioxins and furans containing different proportions of each isomer, toxicity equivalency factors (TEFs), relating the potency of each congener to 2,3,7,8-TCDD have been derived. As a result, dioxin and furan mixtures are referred to as a concentration of PCDD toxic equivalents (TEQs). Several TEF systems that are specific for 2,3,7,8-TCDD, based on toxic potency, have been developed. The general effects following human exposure at high levels include skin lesions, (e.g., chloracne) and some systemic toxic effects (e.g., abnormal liver function and effects on the endocrine system). These effects are slowly reversible, if exposure is terminated before permanent damage occurs and does not appear to lead to long-term health consequences. Because PCDD/Fs are highly lipophilic compounds, they accumulate in fatty tissues, adipose tissue and breast milk and milk products.

POLYCYCLIC AROMATIC HYDROCARBONS (PAHs)

PAHs are carcinogens produced from the incomplete combustion or thermal decomposition (pyrolysis) of organic material. They contaminate foods through environmental contamination or through different heating applications. Nearly 90% of the total environmental PAHs are formed by combustion of coal, residential furnaces and coke ovens. They contaminate foods by air pollution or are formed during broiling, chargrilling, roasting, frying and smoking operations. PAHs are mostly detected in foods including charcoal-broiled meats, smoked and grilled foods, fats and oils, plant materials, seafood, liquid smokes and beverages. The highest levels of PAHs are found in smoked products from traditional kilns that use smoldering wood or sawdust. In modern smoking plants, it is possible to control the combustion process and obtain a desirable smoke-production temperature. The use of external smoke generators offers the possibility of cleaning the smoke by spraying or filtering before it enters the smoking chamber.

Table 7.4   Carcinogenicity of Polycyclic Aromatic Hydrocarbons

IUPAC Name	Carcinogenecity*
Benzo (a) anthracene	+
Benzo (b) fluoranthene	+
Benzo (a) pyrene	+++
Dibenzo (g, h, i) perylene	++
Indeno (1,2,3-c, d) pyrene	+

* +++ very carcinogenic, ++ carcinogenic, + somewhat carcinogenic

Adapted from Gomaa et al., 1993.

Benzo (*a*) pyrene (carcinogenic)   Benzo (*e*) pyrene (noncarcinogenic)

Figure 7.5   Molecular structures of benzo(a)pyrene and benzo(e)pyrene (from Concon, 1988b, with permission from Marcel Dekker, Inc.).

PAHs show mutagenic and carcinogenic effects on experimental animals. Studies on rats showed that feeding a mixture of these compounds together was more carcinogenic than feeding an equivalent dose alone.

While more than 80 PAHs have been identified in the environment, approximately 60% of those containing 24 or fewer ring carbons have been found to be carcinogenic in mammals (Table 7.4).

The carcinogenicity of benzopyrene depends upon its structure. Figure 7.5 shows the chemical structures of benzo(a)pyrene and benzo(e)pyrene. Carcinogenicity is enhanced when a benzene ring is added to the molecule at a strategic position, as in the case of benzo(a pyrene.

In meat cookery, barbecuing is a secondary source of PAH contamination. Benzo(a)pyrene content in smoked sausages has been detected to be between 0–6.15 g/kg, whereas, in barbecued sausages, benzo(a)pyrene varied between 0.1–86.0 g/kg, depending on the fuel and fire conditions. It was also determined that the level of benzo(a)pyrene had increased with the greater fat content of the barbecued sausages. PAHs are also formed during the canning of protein foods, browning reactions, caramelizing of sugar or the roasting of coffee. It has been postulated that PAHs occur in grilled foods because of melted fat dripping from the food and landing on the heat source. The pyrolysis of the fat results in the formation of PAHs and deposition of the toxin on the food as the smoke rises. It was found that charcoal grilling of meat on a vertical grill in which the melted fat does not reach the coals, or broiling with gas or electricity in which the heat source is above the food, results in minimal contamination of food with PAHs. Pickling and fermentation can also produce PAHs, which means that they are found in soy sauce.

RADIONUCLIDES

Radionuclides, or radioactive isotopes, are elements with unstable nuclei that decay or disintegrate at predictable rates. Radionuclides have existed in foods from the beginning of time. They enter the food chain through soil. Most foods contain 1–10 micro-Curies per gram of potassium-40. Radioactivity also comes from the effluents of nuclear power plants, laboratories, debris from aboveground weapons, testing and nuclear accidents. Radioactive iodine in milk is the primary indicator of whether a public health risk exists. Radioactive iodine has high toxicity and is found in greater abundance after a radioactive incident such as Chernobyl. Iodine-131 appears in milk within 3–4 days after release of nuclear material into the environment and milk reaches consumer markets faster than most other foods with potential for contamination from fallout. Milk is also a major source of calories and nutrients for infants, the group most sensitive to radioactive iodine. Strontium-90 contamination occurs on soil after aboveground tests, when the levels of this radioactive compound also increase in foods. Cesium-137 has also been discovered in soil, especially following weapon tests as well as radioactive fallouts. It, too, contaminates foods.

The radionuclides that are most destructive are those that can penetrate the soft tissue and become a part of an organism's active metabolism. Cesium-137 has these properties. It is rapidly absorbed by the bloodstream and can be distributed in all cells of the body. Its half-life is 27 years. Strontium-90, with a half-life of 28 years, is readily absorbed from the gastrointestinal tract or the lung and accumulated in the bones. Oral intake of this substance showed high incidence of cancers of the bone and leukemia. Children under 10, especially, are at great risk. High levels of Iodine-131 destroy the thyroid glands, thus decreasing thyroid hormone production. Levels of Iodine-131 that damage the thyroid but leave the tissue capable of proliferation lead to hyperplastic cancers. Children are twice as susceptible as adults. Potassium iodide should be administered to minimize the dose to the thyroid in situations where exposures may be greater than 10 rads, However. the isotope has a half-life of only 8.1 days.

TRACE METALS

Most mineral elements found in the body, whether essential or nonessential, have high chemical and biological reactivity, particularly in the forms of ions, radicals, or organic complexes; thus, they can be potentially toxic depending on the dose and other conditions. Many mineral elements are required only in trace amounts for their specific physiological functions, but above certain levels they may become toxic. Therefore, an essential trace element becomes a contaminant when it is found in foodstuffs at above nutritionally desirable levels. This section discusses only nonessential inorganic contaminants such as mercury, cadmium, lead, nickel and arsenic. Other trace elements such as tin, silicon, copper, cobalt, beryllium and molybdenum are also found as food contaminants, but evidence to date suggests that they are not toxicologically significant at the levels in which they are found in foods.

Arsenic

Arsenic is released into the environment from both natural and manmade sources. It is found in soils, rocks and natural waters. Man-made arsenic contamination of the environment can arise from smelting of metals and emissions from coal-fired power plants. In the past, arsenic-containing compounds were widely used as pesticides, herbicides and soil sterilants, and were found in fairly high concentrations in some agricultural soils. The widespread distribution of arsenic in the environment is reflected in the occurrence of this toxicant in most foodstuffs, particularly seafood. Chronic arsenism has been mentioned in certain parts of the world because of the high levels of arsenic in the drinking water. In these cases, the type of poisoning that occurs is goiter or skin cancer. Significant examples of acute arsenic poisonings reported in the 20th century are beer poisoning in Manchester caused by the use of arsenic-contaminated sugar in breweries, and, in Japan, well-water poisoning caused by pollution from discharges of a manufacturing plant and milk poisoning caused by the use of arsenic-contaminated sodium phosphate stabilizer.

The deep waters off Taiwan and the waters of southeast Hungary have also been found to contain toxic concentrations of arsenic. There is also good epidemiological evidence implicating arsenic compounds as causative agents in certain skin and lung cancers. The toxicological significance of arsenic depends on its chemical forms. Elemental arsenic is considered nontoxic. The organic forms of arsenic, such as the methyl-, dimethyl-

and trimethyl-arsenic forms are more toxic than the elemental form. The most common form encountered in foods is pentamelant arsenic. Although this is relatively less toxic than other forms of this element, As^{+5} may be transformed to As^{+3} and to methylated form *in vivo* when it becomes toxic.

Cadmium

Cadmium is an element of concern due to its presence in waste products, primarily sewage sludge, that are land disposed. The extensive techno-logical uses of cadmium have resulted in widespread contamination of the soil, air, water, vegetation and food supply by this metal. Cadmium compounds are used in electroplating metals and alloys, in paints, enamels, glazes, textiles, plastics and many other industrial, household and office products and machineries. Plants vary in their capacity to accumulate cadmium. When exposed to high levels of cadmium in the rooting environment, some root crops (turnips) and leafy vegetables (spinach) contain sufficient cadmium to pose a potential health hazard with their consumption. Acute cadmium poisoning results from foods stored in containers where cadmium is used. Such cases would involve acidic foods or beverages such as lemonade, raspberry gelatin, homemade punch, lemon-flavored iced tea and sherbets, when the foodstuffs are allowed to remain in cooling containers such as cadmium-plated ice cube trays or metal pitchers. Another significant source of cadmium is surface water. Some tap water may be contaminated with cadmium picked up as it passes through the pipes. Polyethylene (black), copper and galvanized iron pipes may contain high levels of cadmium. Surface water may also be contaminated by industrial pollution. The toxic effects of cadmium observed in experimental animals include anemia, hypertension, testicular damage and atrophy, and carcinogenesis.

Mercury

Mercury enters the food chain through atmospheric deposition from coal combustion, smelting and volcanic activity and from the use of some types of pesticides. Large amounts of mercury are released into the environment by several groups of industries. Major mercury users are the chloralkali industry (in electrolytic cells), the pulp and paper industry, where mercury compounds are used as shimicides (anti-sliming agents), and agriculture where uses include seed dressings and sprays. Mercury is converted in sediments on river or lake bottoms into highly toxic methyl mercury compounds. Formation of the more volatile dimethyl mercury is favored at alkaline pH. The less volatile, more methyl form is favored at acid pH. Because much of the mercury pollution ends up in rivers and lakes where it is converted into methyl mercury, contamination of fish (especially swordfish and tuna) with mercury is very significant. High levels of mercury have also been detected in meat and certain dairy products. The heavy usage of mercurials in Japan in the treatment of rice plants resulted in the rice's containing elevated levels of mercury. Elemental mercury has little oral toxicity because of poor absorption. The toxicity of inorganic mercury salts is related to absorbability in the gastrointestinal tract. The insoluble mercurous chloride is relatively nontoxic orally, but the mercuric salts are more toxic and the degree of toxicity depends on the type of anions associated with the metal. For example, the oral LD_{50} for mercuric iodide and mercuric acetate in the rat are 40 mg and 104 mg/kg body weight, respectively; the oral LD_{50} of mercuric iodide is >310 mg/kg body weight. The organic complexes of mercury and the alkyl mercury compounds are more toxic. Methyl mercury appears to be absorbed almost completely into the gastrointestinal tract and, in this form, mercury in the brain attains higher levels than with other forms of the metal. In addition to neurological damage in adult animals, alkyl mercury compounds also damage the reproductive system and fetus, particularly the fetal brain. Methyl mercury crosses the placental barrier, causing teratogenic effects.

Nickel

Nickel contamination in foods generally occurs through industrial pollution, but contamination can also be derived from equipment and utensils used in food processing. It has been determined that nickel from stainless steel strips or containers can be leached out to the extent of 0.13–0.22 ppb when in contact with various foods under cooking conditions for 1 hour. It was also reported that more than 400 ppm of nickel may be leached out by acid foods from certain types of stainless steel pans. A nickel alloy catalyst, as used in hydrogenation of oils, may also be a significant source of metal in foodstuffs. The presence of high concentrations of nickel in specific tissue may be responsible for some adverse biological effects. For example, the localization of nickel in the pituitary gland may cause the depression of prolactin secretion under basal conditions. Localization in the pancreas may also result in inhibition of insulin secretion. Nickel carbonyl has elicited more toxicological and public health interest than any other nickel compound. It is a volatile liquid that has many industrial uses, but has no relation to food toxicology. Nickel has carcinogenic potential and the most common form of cancer associated with nickel exposure is cancer of the lungs and nasal cavities, which occur usually by inhalation of the compound.

Lead

Lead, one of the better-known toxic heavy metals, is a major pollutant. Contamination of foodstuffs and surface water derived from both natural and manmade sources appears to be inevitable. Lead is widely used in the manufacture of consumer products such as batteries, paints, gasoline, additives, alloys, glazes, water pipes, insecticides and others. A significant amount of lead comes from automobiles' combustion of gasoline containing the anti-knock agent tetraethyl lead. It is reported that the lead content of vegetable foodstuffs obtained near busy highways is significantly high. Discharges from industrial operations also provide a significant fraction of environmental lead contamination. Lead is introduced into the food chain by deposition on crop plants and soil dust inhalation. Lead contamination of foods also occurs when water flows through lead pipes and glazed earthenware are used as food containers. Lead affects mainly the nervous system, blood, the gastrointestinal tract and the kidneys. The central nervous system is highly susceptible to the toxic effects of lead. Both acute and chronic encephalopathy have been reported. Lead passes both the placental and the blood–brain barriers. Anemia is always present in lead poisoning. The hematological effects are often associated with acute abdominal colic. Damage to the kidneys from acute lead intoxication is also characterized by progressive and possibly irreversible renal failure. Lead also affects thyroid, hypothalamus and adrenal glands and the cardiac muscles. Long-term ingestion of lead acetate in the diet results in the formation of tumors of the lungs, testes, and adrenal, pituitary, prostate, and thyroid glands in rats.

8

FOOD ADDITIVES

Man has a long history of adding chemical substances to foods for various reasons, including improving shelf life, flavor, appearance or texture. Two preservatives, smoke and salt, which were also used to give food a good taste, are probably the oldest food additives. Like spices, they have been used since early times to improve the taste of foods obtained from plant and animal sources. With rapid urbanization in the 19th century, additives became increasingly necessary, especially to protect food against spoilage. In the 20th century, as the industrial production of food developed further, the importance of additives increased — for example, during this time, emulsifiers were developed that improved and simplified the production of margarine. Additives, like trace elements, vitamins and essential amino acids, have also become important. More recently, the importance of flavor enhancers as well as flavoring compounds has increased substantially.

FUNCTIONS OF FOOD ADDITIVES

Food additives are used in food processing to impart many desirable functional properties. They allow our growing urban population to enjoy a variety of safe, wholesome and delicious foods and make possible an array of convenience foods without the inconvenience of shopping daily. Generally, improved keeping and sensory quality can be listed as the main functional effects of food additives. The use of chemical preservatives is one of the most popular and economical methods of improving keeping quality. In the early periods of civilization, wood smoke was used to preserve meats. Today, additives preserve foods against bacteria, molds, yeasts and even viruses. In addition to preservatives, antioxidants make up a group of substances that enhance the keeping quality and stability of foods by preventing rancidity and enzymatic browning reactions, whereas sequestrants bind metals to prevent catalyzing of some food components. The sensory quality of foods is improved by using substances that enhance, change or maintain the taste, aroma, flavor, texture or appearance of foods. Sweeteners and flavorings enhance the taste, aroma and flavor character of foods, whereas coloring agents improve appearance. Emulsifiers, thickeners, anti-caking agents and humectants can be given as examples of the group of additives that improve textural quality. Still more additives are used to provide necessary ingredients or constituents for foods manufactured for groups of consumers having special dietary needs, for example, the addition of nonnutritive sweeteners in foods prepared for dietetic and diabetic purposes.

Many substances that are added to food may seem foreign when listed on an ingredient label, but are actually quite familiar. For example, riboflavin is another name for vitamin B_1; ascorbic acid is vitamin C; alpha-tocopherol is vitamin E; and β-carotene is a source of vitamin A. Although there are no easy synonyms for all additives, it is helpful to remember that all food is made up of chemicals. Carbon, hydrogen and other chemical elements provide the basic building blocks for everything in life.

The Joint FAO/WHO–CAC has issued a table outlining the functional classes, definitions and technological functions of food additives (Table 8.1). The 23 class titles given in this table have been endorsed by the Codex Committee on Food Labeling and were adopted by the 19th session of the commission. A single food additive can often be used for a range of technological functions in a food and it remains to the responsibility of the manufacturer to declare the most descriptive class in the list of ingredients. The list does not include flavors because the Codex General Standard for Labeling does not require these to be specifically identified in the list of ingredients. Further, it does not include chewing gum bases and dietetic and nutritive additives.

LEGAL ASPECTS OF FOOD ADDITIVES

Because food additives are intentional xenobiotics added to foods for various purposes, most countries have regulated the use of these substances in foods. In the U.S., food additives come under the jurisdiction of the Food and Drug Administration (FDA) and the U.S. Department of Agriculture (USDA). The Food Additives Amendment to the Food, Drug and Cosmetic Act of 1938 prohibits the use of food additives as legally defined until its sponsor or proposed user has established its safety. The FDA is mandated to issue regulations specifying the condition of use. The law also exempts from food additive classification many substances generally regarded as safe (the so-called GRAS substances) because of prolonged continued use with no known harmful effects, or affirmed as safe

Table 8.1   Functional Classes, Definitions and Technological Functions of Food Additives

Functional Classes	Definition	Subclasses
Acids	Increase acidity or impart a sour taste to the food	Acidifier
Acidity regulators	Alter or control the acidity or alkalinity of a food	Acid, alkali, base, buffer, buffering agent, pH adjusting agent
Anti-caking agents	Reduce the tendency of particles of food to adhere to one another	Anti-caking agent, anti-sticking agent, drying agent, dusting powder, release agent
Antifoaming agents	Prevent or reduce foaming	Antifoaming agent
Antioxidants	Prolong shelf life of food by protecting against deterioration caused by oxidation, e.g., fat rancidity and color changes	Antioxidant, antioxidant synergist, sequestrant
Bulking agents	Substance, other than air or water that contributes to the bulk of a food without contributing significantly to its available energy value	Bulking agent, filler
Colors	Add or restore color in a food	Color
Color retention agents	Stabilize, retain or intensify the color of a food	Color fixative, color stabilizer
Emulsifiers	Form or maintain a uniform mixture of two or more immiscible phases such as oil and water in a food	Emulsifier, plasticizer, dispersing agent, surfactant, wetting agent
Emulsifying salts	Rearrange cheese proteins in the manufacture of processed cheese, in order to prevent fat separation	Melding salt, sequestrant
Firming agents	Make or keep tissues of fruit or vegetables firm and crisp, or interact with gelling agents to produce or strengthen a gel	Firming agent
Flavor enhancers	Enhance the existing taste or odor of a food	Flavor enhancer, flavor modifier, tenderizer

Table 8.1 *(Continued)* **Functional Classes, Definitions and Technological Functions of Food Additives**

Functional Classes	Definition	Subclasses
Flour treatment agents	Substance added to flour to improve baking quality or color	Bleaching agent, dough improver, flour improver
Foaming agent	Facilitate forming or maintaining uniform dispersion of gaseous phase in liquid/solid food	Whipping agent, aerating agent
Gelling agents	Give food texture through formation of a gel	Gelling agent
Glazing agents	Substances that, when applied to external surface of a food, impart a shiny appearance or provide a protective coating	Coating, sealing agent, polish
Humectants	Prevent foods from drying out by counteracting the effect of a wetting agent atmosphere having a low degree of humidity	Moisture- or water-retention agent, wetting agent
Preservatives	Prolong the shelf life of a food by protecting against deterioration caused by microorganisms	Antimicrobial preservative, antimycotic agent, bacteriophage control agent, chemosterilant and wine-maturing agent, disinfection agent
Propellants	A gas, other than air, that expels a food from a container	Propellant
Raising agent	Substance or combination of substances that liberate gas and thereby increase the volume of dough	Leavening agent, raising agent
Stabilizers	Facilitate maintaining a uniform dispersion of two or more immiscible substances in a food	Binder, firming agent, moisture- or water-retention agent, foam stabilizer
Sweeteners	Non-sugar substance that imparts a sweet taste	Sweetener, artificial or nutritive sweetener
Thickeners	Increase the viscosity of a food	Thickening agent, texturizer, bodying agent

From CAC, 1992, with permission of the Food and Agriculture Organization of the United Nations.

as judged by qualified scientists. The law also contains the Delaney clause, which prohibits the use of food additives shown to cause cancer in humans and animals. In 1985, an amendment was made to the Delaney clause using the *de minimus* concept. This concept is used when a substance is known to cause cancer in laboratory animals but the dietary risk is deemed negligible, so the substance is allowed in foods. For regulatory purposes, negligible means that the lifetime risk of cancer is increased by no more than 1 in a million. The Color Additive Amendment, which became effective in 1960, regulates all colors and requires the reevaluation of the safety of all colors, even those previously considered safe, by means of new scientific tests. It mandates the FDA to set limits on the amounts of colorants used in foods and prohibits the use of colors for such purposes as being deceptive to consumers.

The CAC, within the framework of the Joint FAO/WHO Food Standards Programme, enforces the regulation of food additives at the international level. The commission carries out its work with the help of a number of subsidiary bodies such as the Codex Committee on Food Additives and Contaminants (CCFAC) and the Joint Expert Committee on Food Additives (JECFA). CAC defines a food additive as,

> "any substance not normally consumed as a food by itself and not normally used as a typical ingredient of the food, whether or not it has a nutritive value, the intentional addition of which to food for a technological (including organoleptic) purpose in the manufacture, processing, preparation, treatment, packing, packaging, transport or holding of such food results, or may be reasonably expected to result (directly or indirectly), in it or its by-products becoming a component of or otherwise affecting the characteristics of such foods."

The term does not include contaminants or substances added to food for maintaining or improving nutritional quality.

CCFAC has established a number of principles on the use of food additives that involve the need for their toxicological evaluation and justification, main purposes of use and the requirements for approval or temporary approval to be included in an advisory list or in a food standard. The list of additives evaluated by JECFA and CCFAC is used extensively in setting up or updating national food control systems that include control of food additives. An international numbering system (INS) has also been prepared by CCFAC for the purpose of providing an agreed-upon international numerical system in ingredient lists as an alternative to declaration of a specific name, which is often a lengthy and a complex chemical structure. This system was based on the restricted system already introduced successfully within the European Community (EC).

In Europe, the formation of the EC has created the requirement of bringing food additive approvals of the member nations into alignment, so as to eliminate differences in laws that hinder the movement of foodstuffs among these nations. Historically, the member countries had different approaches to food additive approval and to their tendency to approve new additives. A council directive on the approximation of the laws of the member states concerning food additives is already in use. The definition, classes and principles of use established for food additives in EC directives are nearly the same as those mentioned by CAC.

SAFETY ASPECTS OF FOOD ADDITIVES

Before they are permitted for use in food, additives undergo various testing to ensure their safety. To determine the safety of potential food additives, several types of tests are performed including acute, genetic and pharmacokinetic studies, subchronic studies involving reproduction, and teratology and chronic studies involving mutagenesis and carcinogenesis. Safety evaluation of food additives is a two-stage process. The first stage involves the collection of relevant data, including the results of studies on experimental animals and, where possible, observations in man. The second stage involves the assessment of such data to determine the acceptability of the substance as a food additive. International assessment of food additives is provided by JECFA of the CAC. JECFA comprises two panels covering the two basic aspects of a food additive. While a group of experts look at the chemistry of the compound, the toxicological panel reviews all the available and relevant information on the biochemical aspects of the compound such as its absorption, distribution, biotransformation and effects on enzymes and excretion, then conducts acute short-term and long-term toxicity tests. As a result of these evaluations, the acceptable daily intake (ADI) values for food additives are established. ADI is an estimate by JECFA of the amount of a food additive, expressed on a body weight basis, that can be ingested daily over a lifetime without appreciable health risk (standard man = 60 kg). JECFA generally sets the ADI of a food additive on the basis of the highest NOAEL in animal studies. For substances that induce a toxic response (other than cancer) in chronic feeding tests, the NOAEL is used to determine the ADI values. Because ADI value is derived from the NOAEL, usually from animal studies, appropriate empirical or data-derived safety (uncertainty) factors are applied in calculating this value.

$$ADI = \frac{NOAEL}{Uncertainty\ factor}$$

In the formula given above, NOAEL value is calculated in mg/kg body weight/day, determined by proper selection of the doses of a chemical in an animal study, such that the highest dose produces an adverse effect that is not observed at the lowest dose. When results from two or more animal studies are present, the ADI value should be based on the most sensitive animal species that displayed the toxic effect at the lowest dose. A 100-fold uncertainty factor (safety factor) is widely used, but may be modified when adequate human data are available or based on comparative pharmacokinetic/toxicodynamic data. The resulting figure (ADI) expressed on the basis of mg of the chemical per kg of body weight is related to the daily lifetime consumption of the additive without apparent health risk. If the food additive is going to be used in an infant formula (below the age of 12 weeks), the toxicity studies should be conducted on very young animals as well. In cases where infants more than 12 weeks of age may represent a risk group based on the toxicokinetic and toxicodynamic assessments, risk management procedure should be employed to confirm that the ADI values are not exceeded. ADI "not specified" (NS) is a term applicable to a food substance of very low toxicity which, based on available data (chemical, biochemical, toxicological and other), does not, in the opinion of JECFA, represent a hazard to health. JECFA bases this acceptance on the total dietary intake of the substance at levels necessary to achieve the desired effect and from its acceptable background in foods. For that reason and for reasons stated in individual JECFA evaluations, establishment of an ADI expressed in numerical form is not deemed necessary by JECFA. JECFA leaves it to national governments to regulate food additives so that consumption from natural occurrence plus the deliberate addition of food does not exceed the ADI for each additive that is permitted. The ADI NS food additives are generally regarded as safe at normal levels of use and can be added to foods at good manufacturing practice (GMP) levels in most countries.

GENERAL PRINCIPLES ON THE USE OF FOOD ADDITIVES

1. Only those food additives shall be endorsed and included in Codex General Standard for Food Additives which, so far as can be judged on the evidence presently available from JECFA, present no risk to the health of the consumer at the levels of use proposed.
2. The inclusion of a food additive in Codex General Standard for Food Additives shall have taken into account any acceptable daily

intake, or equivalent assessment, established for the additive and its probable daily intake from all sources. Where the food additive is to be used in foods eaten by special groups of consumers, consumers in those groups shall take account of the probable daily intake of the food additive.

3. The use of food additives is justified only when such use has an advantage, does not present a hazard to health and serves one or more of the technological functions set out by Codex, and needs set out from (a) through (d) below and only where these objectives cannot be achieved by other means that are economically and technologically practicable:

   a. To preserve the nutritional quality of the food; an intentional reduction in the nutritional quality of a food would be justified in the circumstances dealt with in sub-paragraph (b) and also in other circumstances where the food does not constitute a significant item in a normal diet.

   b. To provide necessary ingredients or constituents for foods manufactured for groups of consumers having special needs

   c. To enhance the keeping quality or stability of a food or to improve its organoleptic properties, provided that is does not change the nature, substances or quality of the food so as to deceive the consumer.

   d. To provide aids in the manufacture, processing, preparation, treatment, packing, transport or storage of food, provided that the additive is not used to disguise the effects of the use of faulty raw materials or of undesirable (including unhygienic) practices of techniques during the course of any of these activities.

4. Good manufacturing practice (GMP): All food additives subject to the provisions of Codex General Standard for Food Additives shall be under conditions of good manufacturing practice, which include the following:

   a. The quantity of the additive added to food shall be limited to the lowest possible level necessary to accomplish its desired effect.

   b. The quantity of the food additive that becomes a component of food as a result of its use in the manufacturing, processing or packaging of a food and which is not intended to accomplish any physical, or other technical effect in the food itself, is reduced to the extent reasonably possible.

   c. The additive is prepared and handled in the same way as a food ingredient.

5. Specifications for the identity and purity of food additives: Food additives used in accordance with Codex General Standard for

Food Additives should be of appropriate food grade quality and should at all times conform with the applicable Specifications for Identity and Purity recommended by the Codex Alimentarius Commission or, in the absence of such specifications, with appropriate specifications developed by responsible national or international bodies. In terms of safety, food grade quality is achieved by compliance with the specifications as a whole and not merely with individual criteria. — *From CAC, 1997, with permission of the Food and Agriculture Organization of the United Nations.*

COMMON USES OF ADDITIVES

Different types of additives are used for different purposes, though many individual additives perform more than one function. For the purposes of both classification and regulation, they are grouped according to their primary functions. The main groupings, or classes, of additives are explained below, together with their functions and some examples of their use.

Acids

Acids (acidulants) are used to give a sour taste to foods in a controlled way. The taste of all acidulants is usually described as "tart." However, each individual acid has a slightly different tartness. Among the organic acids, the tartness of citric acid has been described as clean; that of malic acid as smooth; fumaric acid as metallic; adipic acid as chalky; acetic acid as astringent; tartaric acid as sharp or bitter; and lactic acid as sour. Phosphoric acid, which is inorganic, tends to have a flat sourness. Besides being sour, most of these acids have a characteristic taste of their own and complement fruit and other flavors in carbonated beverages, preserves, fruit drinks and desserts. For example, butyric acid in low concentrations contributes to the typical flavor of such products as cheese and butter. Acids are also used to modify sweetness and enhance flavor in some foods. They add the tartness required to balance the excessive sweetness of such products as hard candies, gelatin desserts, carbonated and noncarbonated beverages, jellies, preserves, toppings, etc. The sodium salts of gluconic, acetic, citric and phosphoric acids are commonly used for pH control and tartness modification by the food industry. The ability of acids to lower pH makes them useful as preservatives because an acidic medium retards the growth of the microorganisms responsible for spoilage, as well as preventing enzymatic browning in fruits. Acids also aid in sterilization of foods and many food products that formerly could not have been adequately sterilized. Such foods can be safely processed and maintained for long periods of time through the technique of acidification. For example, acids like citric are added to some moderately acid fruits and vegetables to lower the pH to a value below 4.5. In canned foods, this permits sterilization to be achieved under less severe thermal conditions than is necessary for less acid products and has the additional advantage of preventing the growth of hazardous microorganisms such as *Clostridium botulinum*.

Acidity Regulators

Acidity regulators are added to change or maintain the acidity or alkalinity of foods. They include buffers, acids, neutralizing agents and peeling agents (alkalis such as sodium hydroxide and potassium hydroxide) for tomatoes, peaches, potatoes, pears and cocoa products. Sodium acetate is a buffer used in breakfast cereals and baked goods. Acids include acetic acid in sherbets, syrups and beverages and fumaric acid in confectionery. Citric, malic and tartaric acids impart a tart taste to soft drinks. Buffering agents, which are generally the sodium salts of these acids, are frequently used to control the degree of acidity in soft drinks. Concentrations of such acids and buffers are essentially the same as the levels at which these substances occur in natural fruits. In cola-type beverages, the most commonly used acidity regulator is phosphoric acid. Alkalis include magnesium hydroxide in canned peas; calcium oxide in ice cream mixes, sour cream and butter; and sodium bicarbonate in baking mixes, tomato soup and cocoa products. Many foods, such as cucumbers, cannot be preserved unless they are made acidic as in the form of pickles. Phosphates, acetates and carbonates are buffers that stabilize the pH of a product. Adjustment of acidity is necessary, for instance, in the production and use of several dairy products. Excessive acidity that may develop in ice cream must be neutralized for satisfactory churning and to produce a butter of acceptable flavor and keeping quality. Emulsification and a desired tartness in processed cheese and cheese spreads are obtained by the addition of citric, lactic, malic, tartaric and phosphoric acids.

Anticaking Agents

Anticaking agents are substances that are insoluble in water and have the capacity to absorb water. They are used to maintain the free-flowing characteristics of granular and powdered forms of hygroscopic foods. They function by absorbing excess moisture, or by coating particles and making them water repellent. Calcium silicate, used to prevent caking in baking powder, table salt and other food products, absorbs oil in addition to water and can be used in powdered mixes and spices that contain free oils. Calcium and magnesium salts of long-chain fatty acids such as calcium stearate are used as anticaking agents in dehydrated vegetable products, salt, onion and garlic powders. Other anticaking agents employed in the food industry include sodium silicoaluminate, tricalcium phosphate, magnesium silicate and magnesium carbonate.

Antifoaming Agents

Foaming, a frequent problem in food manufacturing operations, causes production inefficiencies. Polydimethylsiloxane, or silicone, is used at a level of approximately 10 parts per million to control foam of products. The silicone disperses itself throughout the liquid film that makes up the foam and causes it to collapse.

Antioxidants

Antioxidants interfere with the early stages of oxidative and auto-oxidative processes to prevent formation of unwanted reaction products. These substances delay oxidative spoilage of foods and reduce the oxidative deterioration that leads to rancidity, off-flavors, off-odors and loss of color and nutritive value of foods. The rate of deterioration, which can vary considerably in different foods, is influenced by the presence of natural antioxidants and other components, availability of oxygen and sensitivity of the substance to oxidation, temperature and light. Although oxidation can be prevented or retarded by replacing air with inert packaging gases, removal of oxygen with glucose oxidase, incorporation of UV-absorbing substances in transparent packaging materials, or cooling and use of sequestrants, these treatments may not possible in all cases, or, indeed, sufficient for various types of foods. Thus, antioxidants, e.g., ascorbic acid, are used to retard oxidative deterioration, while others interfere in the mechanism of oxidation, e.g., tocopherols, gallic acid esters (dodecyl, octyly and propyl gallates), butylated hydroxyanisole (BHA) and butylated hydroxytoluene (BHT). All have specific properties that make them more effective in some applications than in others. The use of the powerful synthetic antioxidants BHA and BHT and the gallic acid esters is very restricted. Tocopherols, which can be either natural or synthetic, are less restricted, but are less effective in the protection of processed foods. Antioxidants can also prevent unwanted oxidative changes in essential oils, aroma compounds and chewing gums. Fruit and potato-based products, in particular, can be protected against discoloration through addition of ascorbic acid, sulfur dioxide and sulfites. By reducing the availability of metallic ions that may catalyze oxidation reactions, often a combination of two or more antioxidants is more effective than any one simply because of their synergistic effect, for example, isoascorbic acid (erythorbic acid) and salts. To optimize this synergism, the necessary ratio of the two antioxidants must be established experimentally. In addition, some anti-oxidants inhibit lipoxygenase and, by this side effect, can prevent the enzymatic oxidation of fats.

Antioxidants have the desired effect only within a certain concentration range. At excessive concentration, they may act as pro-oxidants. If the aim is to protect fats in the finished food product after it has been heated, antioxidants that maintain their effectiveness after such processing must be used. It should be noted that antioxidants cannot restore oxidized food; they can only retard the oxidation process. As oxidation is a chain reaction process, it needs to be retarded as early as possible. Antioxidants, which must be introduced into food prior to the onset of oxidation, are the most effective means of preventing or reducing the oxidation problem and enhancing the keeping quality of foods.

Bulking Agents

Bulking agents are substances that add volume to food products without contributing to their caloric value. In applications where sugar is replaced by a high-intensity sweetener, bulking agents make up for the lost volume and ideally provide some or all of the functional properties of sugar. The most important properties of a bulking agent are reduced calorie content through limited digestibility, solubility and minimal side effects. Polydex-trose, a polymer of glucose that contains sorbitol and citric acid, is the most widely used bulking agent in bakery products, frozen desserts, candy and confectionery products, jams, chewing gums, salad dressings, gelatins and puddings.

Coloring Agents

Color is the first characteristic of food that is noticed, and this predeter-mines the expectation of both flavor and quality. Coloring agents have been used in foods and drinks for centuries. Until the discovery of dye synthesis in 1856, natural extracts from animal, vegetable and mineral origin were used. These included annatto, saffron, turmeric and caramel. In 1856, Perkin produced the first synthetic dye — mauve or mauveine — through the oxidation of crude aniline. Since that breakthrough, the use of artificial colors has increased; until today, about 90% of all colors are synthetic. Generally, colors can be classified as natural or synthetic according to their origin. Synthetic colors may be obtained as the nature-identical forms of natural colors, or as artificial colors that have no resemblance to the natural ones. Natural colors can be of both organic or inorganic origin. Natural organic colors are derived from natural edible sources using recognized food preparation methods. The typical examples of this group are anthocyanins; beetroot, which contains the pigments collectively known as betalains; cochineal or carminic acid; curcumin; chlorophyll; annatto and β-carotene; paprika; riboflavin; chlorophyll and copper chlorophyll; caramel; ferrous gluconate and carbon black. The most widely used natural inorganic colors are titanium dioxide, iron oxides and hydroxides, and calcium carbonate. Artificial organic colors were also known as "coal tar dyes" because, for about 80 years following the synthesis of mauve, coal tar was the principal source of food colors.

Most artificial colors are water-soluble and are insoluble in most organic solvents. They have high tinctorial power, so only exceedingly small amounts are needed to color foods. In the U.S., these are designated "certified colors." From a regulatory standpoint in the U.S., with the passage of the 1938 Federal Food, Drug and Cosmetic Act, certification is given to food colorants by FDA to assure that they present no hazard to the health of consumers when they are used at permitted levels in foods. These colors are referred to as FD&C-certified colors, indicating that they are certified by the Food and Drug Administration to be used in foods. They are all water-soluble dyes, but can be transformed into insoluble pigments known as "lake colors" by precipitating the dyes with aluminum, calcium, or magnesium salts on a substrate of aluminum hydroxide. Stability is satisfactory in the 3.5–9.5-pH range. They are used for coloring oil- and fat-based mixes such as biscuits, cakes, icings, fillings, chocolate substances, salad dressings, snack foods, soups and spices where the presence of water is undesirable. The artificial colors which are permitted by the European Community (EC) are given an E number, indicating that they are natural and nature-identical colorings. Typical examples of artificial colors used in foods are tartrazine (yellow), sunset yellow FCF (yellow orange), quinoline yellow (bright greenish yellow), allura red (red), ponceau 4R (bright red), amaranth (bluish red), erythrosine (bright bluish red), carmoisine or azorubine (bluish red), red 2G (bluish red), patent blue V (bright blue), indigo carmine or indigotine (blue), green S (bluish green), brilliant blue FCF (bright greenish blue), chocolate brown HT (reddish brown), brown FK (yellowish brown) and brilliant black BN or black BN (violet).

Color Retention Agents

Color retention agents, which are also known as color fixatives or color stabilizers, help foods retain their natural color during processing and storage, and prevent undesirable discoloration. These additives are especially important with meat products. The red pigment myoglobin can be oxidized by oxygen in the air to form metmyoglobin, which is brown to grayish brown. Nitrates or nitrites stabilize the desired meat color through formation of nitrosomyoglobin. This complex has good storage, cooling and boiling and baking stability.

Emulsifiers

Emulsifiers are substances that are used to facilitate the mixing together of ingredients that normally would not mix, such as oil and water. An emulsifier consists of a hydrophobic and a hydrophilic group, thus enabling the combining of two or more immiscible phases. Emulsifiers improve and stabilize the consistency of foods and sometimes their viscosity, consistency, texture and mouthfeel, as well. They improve the shelf life of some foods, such as baked products. Emulsifiers are essential in the production of mayonnaise, chocolate products and fat spreads. They are also used in ice cream manufacture, yeast-leavened products, shortenings and cured-meat products such as sausages.

The most popular emulsifiers used in foods are lecithin, phosphates, diphosphates, triphosphates, polyphosphates, sodium, potassium and calcium salts of fatty acids, mono- and diglycerides of fatty acids, esters of mono- and diglycerides of fatty acids, sucrose esters of fatty acids, sucroglycerides, polyglycerol esters of fatty acids, propane-1, 2-diol esters of fatty acids, sodium and calcium stearoyl-2-lactylate and stearyl tartrate.

Emulsifying Salts

Emulsifying salts are substances that, in the manufacture of processed cheese, rearrange cheese proteins in order to prevent fat separation. They inactivate calcium, which is important for the stability of the cheese gel. The most common emulsifying salts in foods are citrates and ortho-, di-, or polyphophates.

Firming Agents

Firming agents are used to retain the textural quality of fruits and vegetables. During thermal processing and freezing, the bonds of specific substances in plant walls that help to stabilize structure are modified, resulting in an unacceptably soft product. Adding polyvalent cations, which promote the cross-linking of the free-carboxyl groups of pectic substances, can strengthen the cell wall structure of fruits and vegetables. Addition of calcium salts prior to processing may firm fruits such as tomatoes, berries and apple slices. The most common calcium salts used are calcium chloride, calcium citrate, calcium sulfate, calcium lactate and monocalcium phosphate. Acidic aluminum salts such as sodium aluminum sulfate, potassium aluminum sulfate, ammonium aluminum sulfate and aluminum sulfates are added during the preparation of pickles to provide firming effect.

Flavorings

Flavor (or flavoring) is the property of a substance that causes a seemingly simultaneous sensation of taste in the mouth and odor in the back of the nose. The flavor property of a substance can be moved or transferred to other substrates, such as foods and beverages. This characteristic is what the flavorist (and the flavor industry) rely upon when creating a flavor composition destined for a specific application. Cinnamon, cloves, pepper and fruit extracts have long been used. Today, with the increase in foreign travel and spread of world cultures, food companies have become increasingly conscious of the vital importance of high-quality flavors in the public acceptance of foods. Flavorings are not categorized in the INS system of CAC because the Codex General Standard for Labeling does not require these to be specifically identified in the list of ingredients. In the EC flavor directive, flavorings are defined in the following categories:

Flavoring Substances

These are classified in three groups according to their method of production processes as natural, nature-identical or artificial flavorings. **Natural flavorings** are obtained through purely physical processes, usually from plant material or through microbiological processes. Spices and herbs are considered to be the first natural flavorings utilized by man. In addition to their flavors, they were also employed for their preserving effect on meat products. However, they have poor flavor strength and there is batch-to-batch variation in their flavoring potentials. They have other disadvantages, such as low stability and being often contaminated with bacteria and mold spores. In view of the disadvantages encountered in using spices and herbs, flavorists prefer corresponding essential oils and oleoresins.

Essential oils are aromatic, or odorous, oily liquids (sometimes semi-liquid or solid) obtained from plant material. The oils volatilize, that is, evaporate, from the botanical upon heating and it is their volatility that distinguishes them from fatty oils. The essential oils are derived from specific parts of plants such as flowers, buds, seeds, fruits, leaves, twigs, bark and roots and they make up the most flavorful part of the plant.

Oleoresins are prepared by percolating a volatile solvent, such as a chlorinated hydrocarbon, through a ground spice or herb. They differ from essential oils in that they contain many of high-boiling and nonvolatile constituents native to the spice and herb and can also be used for their coloring potential as well as for their flavor value.

Isolates are aromatic chemicals derived from natural products. The most frequently used methods for obtaining isolates from essential oils are distillation, crystallization and extraction. When a naturally occurring flavor material is not prepared synthetically, it is usually because it is easier and more economical to isolate it from the natural source, However, because the production of isolates is strictly dependent on the availability of the parent essential oil or other natural raw material, industry has moved more toward their production via synthesis.

Extracts are solutions obtained by passing alcohol, or an alcohol–water mixture through a substance. Vanilla extracts and fruit juices can be given as examples. To overcome the weak flavor strength of fruit extracts, the flavor trade has developed reinforced extracts, commonly referred as WONF (with other natural flavors) extract. A WONF extract is a fruit extract or other natural flavor containing both a characterizing flavor from the product whose flavor is being simulated, and the reinforced extract to reinforce the characteristic flavor. Juice extracts can be given as examples.

Nature-identical flavorings are produced synthetically, but are identical to their natural counterparts. When used in comparable concentrations in foods they cannot differ in their action or their safety from the equivalent natural product. They are organic chemicals obtained by synthesis or isolated by chemical means from natural raw materials. These substances are chemically identical to substances present in natural products either as such or processed for human consumption (e.g., vanillin from wood lignin equates with vanillin in vanilla beans). To obtain nature-identical flavorings, the structures of natural substances are identified by sophisticated chromatographic and spectroscopic techniques and synthesized in the laboratory.

Artificial flavorings are substances synthetically produced that have not been found in any natural products suitable for human consumption. The definition of artificial as used by CAC is "those substances that have not yet been identified in natural products intended for human consumption, either processed or not." These substances are particularly appropriate for foods exposed to severe processing conditions. Two common examples are allyl hexanoate and aldehyde C-16. In some countries, only artificial flavorings are considered additives within the meaning of food law.

Flavoring Preparations

According to the EC flavor directive, these are substances with flavoring properties that are obtained by appropriate physical means (including distillation and solvent extraction) or by enzymatic or microbial processes, from material of vegetable or animal origin, either in the raw state or after processing for human consumption by traditional food preparation processes (including drying, torrefaction and fermentation). Essential absolutes, meat extracts and enzyme modifiers can be given as examples of this group.

Process Flavorings

This group of flavorings is obtained according to good manufacturing practices by heating a mixture of ingredients, not necessarily themselves having flavoring properties, of which at least one contains nitrogen (amino) and another is a reducing sugar, to a temperature not exceeding 180°C for a period not exceeding 15 minutes. The most important examples of this category are meat flavorings.

Smoke Flavorings

These are used in traditional smoking processes. The 3,4-benzopyrene content coming from these flavorings is limited to 0.03 µg/kg in foodstuffs and beverages.

Flavor Enhancers

Flavor enhancers can be defined as substances that enhance the existing taste or odor of a food. These substances do not have flavor themselves in the low levels used, but intensify the flavor of other compounds present in foods. Early in the 1900s, the Japanese discovered that the glutamate of seaweed was responsible for its flavor-enhancing properties. Since then, glutamates have been utilized for years as flavor enhancers in one form or another, for example, as crystalline salts such as monosodium glutamate (MSG) or as plant protein hydrolyzates. The best-known and most widely used flavor enhancer is MSG. Its flavor-enhancing properties were discovered by Japanese scientist Dr. Ikeda in 1908. MSG is used in preparation of frozen foods containing meat and fish, dry soup mixes and canned foods. It is not flavorless and has a taste described by the word *umami*, derived from the Japanese for deliciousness, but it doesn't add any flavor in the concentrations used. It is effective in enhancing the flavor of foods in parts per thousand. There are compounds, however, that are significantly effective in enhancing the flavor of foods in concentrations of parts per billion and they are known as 5'-nucleotides. Disodium-5'-inosinate and disodium-5'-guanylate can be given as examples of 5'-nucleotides, which act well with most foods that are enhanced by MSG. A different type of flavor enhancer is maltol, which has the ability to enhance sweetness produced by sugars. It is particularly effective in modifying or intensifying the flavor of soft drinks, fruit drinks, jams, gelatin and other foods high in carbohydrates. CAC also permits the use of ethyl maltol, potassium, calcium, magnesium glutamates, potassium and calcium salts of inosinic and guanylic acids as flavor enhancers in foods.

Flour Treatment Agents

Flour treatment agents can be used to inhibit discoloration in light-colored foods. Freshly milled wheat flour has a pale yellow tint and yields sticky dough that does not handle or bake well. Both the color and baking properties improve slowly in normal storage. These improvements can be obtained more rapidly and with better control through the use of certain oxidizing agents. Benzoyl peroxide is an agent that oxidizes the carotenoid pigments, resulting in white flour. Oxides of nitrogen, chlorine dioxide and other chlorine compounds both bleach color and mature the flour.

Foaming Agents

Foaming agents are used to form or maintain uniform dispersion of a gas in a liquid or solid food. Whipping of egg white, raising of bread flour can be given as examples. Foaming agents are mainly used in bakery and confectionery products. The examples are methyl ethyl cellulose, fatty acids and triethyl citrate.

Gelling Agents

Gelling agents combine with water to form pseudogels or gels. They maintain or improve the structure, consistency or elasticity of a food. Gelling agents facilitate development of the gel structure desired in some foods. They are used in confectionery, desserts, jams, preserves and coating for fish and meat products. They also increase the freeze–thaw stability of frozen foods. The action of some gelling agents depends on the calcium ion content and pH of the food. The most important gelling agents used in food are gelatin, alginates, agar, carrageenan and pectin.

Glazing Agents

Glazing agents are substances that provide a shiny appearance or a protective coating when they are applied to the external surface of a food. These substances are used to protect foods or their surfaces against undesirable changes such as drying out and loss of aroma. They are also added to prevent the growth of undesirable microorganisms on the surface of the food. Some glazing agents are used for enhancing the attractive appearance of foods. The most important glazing agents used in food are waxes, resins, oils, cellulose esters or acetic acid esters of the monoglycerides of edible fatty acids and talc.

Humectants

Humectants are hygroscopic substances added to foods to retain moisture. They are added to food to maintain a predetermined moisture level. They thereby prevent excessive drying out, or any changes in texture associated with this and hardening. In intermediate-moisture food intended to have a long shelf life without expensive packaging, the role of humectants is to standardize water activity to 0.85 or less. This prevents the multiplication of bacteria, which require water activity of 0.90 or more for growth. Growth of molds and yeasts is still possible below this limit, but it can be suppressed easily by addition of sorbic acid. In confectionery, humectants can prevent undesirable crystallization of the sugar. They contribute to the softness or chewiness of confectionery. Sorbitol is used in shredded coconut, frozen desserts, dietetic fruit packs and soft drinks, and glycerol in marshmallows, pastilles and jelly-like candies. Sodium tripolyphosphate is used in sausages to retain moisture.

Preservatives

Preservatives are compounds that delay or prevent microbiological spoilage of foods. They not only act against visible spoilage by yeasts, molds and bacteria, but also prevent the formation of toxins, especially those produced by bacteria and molds. Sugar, salt and wood smoke were used to preserve foods in the days before refrigeration and modern processing techniques. These methods, however, are not compatible with all food products; thus preservatives, also known as antimicrobials, are used. Preservatives are probably the single most important class of additives, because they play an important role in the safety of the food supply. The use of chemical preservatives, such as sulfur dioxide and sulfites, is a continuation of the age-old practices of using salt, sulfite and spices. All raw food materials are subject to biochemical processes and microbiological action that limit their keeping qualities. Preservatives are used to extend the shelf life of certain products and to ensure their safety through that extended period. Most importantly, they retard bacterial degradation, which can lead to the production of toxins and cause food poisoning. Thus, from a toxicological point of view, they offer an important consumer benefit in keeping food safe over the shelf life of the product.

The most widely used preservatives are benzoates, sorbates, propionates, sulfur dioxide and sulfites, parabens, sodium nitrate, sodium nitrite, natamycin and nisin. Preservatives are mainly used in cheese, meat products, fruit-based products, beverages and baked goods. Depending on the type of food and the spoilage expected, different preservatives can be used. For example, meat products are preserved mainly with common salt and nitrite; fruit-based products with sulfur dioxide; beverages with sorbic acid; and baked goods with sorbic acid or propionates. It should be remembered that no preservative is equally effective against all microorganisms. Most substances used today are more effective against yeasts and molds than against bacteria. Nitrites and nitrates are effective against *C. botulinum*, which is the causative organism for the botulinus toxin. Unlike antibiotics, preservatives do not lead to the development of resistant organisms.

Propellants

Propellants are used in the production of foods in aerosol form such as whipping cream; frying, baking and mold-release oils; and spice extracts. Carbon dioxide, carbon monoxide and certain fluorohydrocarbons and mixtures of these are common types of propellants used in foods.

The natural atmospheric gases are also used for preventing oxidative or bacterial spoilage in certain types of prepacked products such as meat, fish and seafoods, fresh pastas and ready-prepared vegetables found on the chilled food counters in sealed containers. The "head space" of the container is filled with one or a combination of gases, depending on the product, to replace the air and modify the atmosphere within the pack to help retard oxidative or bacteriological deterioration, which would occur under normal atmospheric conditions. Nitrogen or carbon dioxide gases are frequently used for these purposes. These gases have no additive function as they are not detectable in the food itself and function only to preserve the food in its packaged state for a longer period, but, for regulatory purposes, they are classified as additives and must therefore be labeled.

Raising Agents

Many bakery products, such as self-rising flours, prepared baking mixes and refrigerated dough, rely on chemical raising agents to produce the gas that gives them volume. Bicarbonates produce carbon dioxide in the presence of heat and moisture. Sodium bicarbonate is the most commonly used product, but ammonium bicarbonate and potassium bicarbonate are used as well. When used alone, sodium bicarbonate reacts to give products a bitter, soapy flavor, thus, it is always combined with a raising agent.

Raising agents are classified according to the rate at which they release carbon dioxide from sodium bicarbonate. Some agents begin producing carbon dioxide as soon as they come into contact with water; others do not begin to react unless heat is present as well. The type of raising agent needed depends on the product. For example, refrigerated dough products require limited carbon dioxide release initially so that they can be packed into containers, but need significant activity upon heating. A slow-acting raising agent would be used for this product. Doughnuts, which must be leavened prior to being exposed to heat, require fast-acting raising agents. Most products use both slow- and fast-acting raising agents to obtain the appropriate volume. The raising agents most frequently used include potassium acid tartrate, sodium aluminum sulfate, δ-gluconolactone and ortho- and pyrophosphates.

Raising agents lighten baked products, mostly by liberating carbon dioxide, which can be generated biologically or chemically. The latter method relies on a mixture of sodium hydrogen carbonate and an acid, e.g., a solid organic acid such as tartaric or citric acid, or sour salts such as certain phosphates.

Stabilizers

Stabilizers are compounds that provide stability for emulsions, suspensions and foams. Stabilize means to hold water in. Most of these materials are hydrocolloids such as gum arabic, guar gum, carrageenan, xanthan gum, locust bean gum, cellulose derivatives such as carboxy methylcellulose, starch, pectin, gelatin and alginates. They are used in baked foods to improve volume, uniformity and fineness of grain; in frozen desserts to control the size of ice crystals and to improve texture; in confectionery products to maintain homogeneity and to improve keeping quality and texture; in beer as foam stabilizers; and in chocolate milk to increase viscosity and prevent the settling of cocoa particles to the bottom of the container.

Sweeteners

The most widespread users of these additives are the low-calorie foods, which, in addition to helping control weight, have also made it possible for diabetics to enjoy several foods they otherwise would not be able to eat. Compounds that provide sweetness without requiring insulin for their metabolism are very useful for individuals with diabetes mellitus. Sweeteners may be nutritive, as the hydrogenated sugars, also known as sugar alcohols or polyols; or nonnutritive, as the intense sweeteners. "Nutritive sweeteners" are composed of sugar alcohols known as polyols, which are produced by the hydrogenation of sugars and syrups with the aid of a catalyst, usually Ranney nickel. Among these, polyols, sorbitol, mannitol and xylitol are naturally present in some fruits and vegetables and can be obtained by the hydrogenation of monosaccharides. Isomalt, maltitol and lactitol are obtained by hydrogenation of disaccharides.

Hydrogenated glucose syrups are mixtures of hydrogenated saccharides and polysaccharides. The sweetening power of polyols is generally lower than or equivalent to that of sucrose. Like glucose or sucrose, they have a bulking effect, which means they can be substituted for traditional sugars. Therefore, they are often called "bulk sweeteners." Their metabolism is like that of slow sugars and therefore leads to a glycemic and insulinemic response lower than that of glucose. As they can be partly digested in the colon, their energy value is lower than other sugars. In the colon, they act like dietary fibers and could facilitate intestinal transit. They may also have a beneficial role in the prevention of digestive cancerous and metabolic diseases. These sweeteners have also been used for products like sugarless gums and mints and as sweeteners for medicines, because they are believed to be less cariogenic (cavity-producing) than other sugars. Unlike sugars, they are not readily fermented and bacteria are unable to convert them into acids. The risk of caries is lower with confectionery made from nutritional sweeteners than with those made from sugars. A drawback to sugar alcohols is their laxative effect when consumed in large doses. However, the EC Scientific Committee on Food (SCF) advised that 20g per person per day of polyols is unlikely to cause undesirable laxative symptoms. In some countries, product labels may be required to carry an appropriate warning, depending upon the amounts involved.

Nonnutritive sweeteners and low-calorie sweeteners involve a broad group of substances that provide a sweet taste or enhance the perception of sweet taste without increasing the caloric value of foods. They are also known as "intense sweeteners" because these are compounds with far more intensely sweet taste than sucrose. Because of their sweetness, they can be added only in small amounts and do not contribute significantly to the caloric value of food. Saccharin has been commercially available to sweeten foods and beverages since around the turn of the century. It is approximately 300 times sweeter than sucrose. It is stable over a wide range of temperatures and conditions. Its GRAS status was removed in 1972 in the U.S. by the results of a study showing a possible link between this sweetener and bladder cancer in rats. Legally, it is now classified as a cocarcinogen (tumor-promoter) with very low potency. Extrapolation suggests that saccharin intake at 30–300 mg per day does not increase human cancer risk; therefore it is allowed in the U.S. and in many countries today. It has been determined as safe by both JECFA and the EC Scientific Committee on Food. Discovered in 1937, cyclamates were 30 times sweeter than sucrose. Due to their relation with bladder cancer in rats, they were banned in 1970 in the U.S. Today, both WHO and EC consider these substances safe.

Aspartame is a dipeptide derivative that was approved in the U.S. in 1981. It is metabolized in the body to phenylalanine, aspartic acid and methanol. Only people with phenylketonuria cannot break down phenylalanine. Therefore, to allow avoidance of consumption by phenylketonurics, aspartame-sweetened products must be labeled prominently regarding their phenylalanine content. Aspartame is a caloric substance because it is a dipeptide that is completely digested after consumption. However, its intense sweetness (about 200 times as sweet as sucrose at a 4% concentration) allows functionality to be achieved at very low levels that provide few calories. The stability of aspartame is dependent on pH, moisture and length of storage. In aqueous systems and at neutral and alkaline pH values, diketopiperizine is formed with the chemical breakdown of the compounds in aspartame, especially when the food is heated or stored at elevated temperatures. A perceptible loss of sweetness occurs with its formation but there is no toxic effect.

Discovered in 1967, acesulfame K is another sweetener that is about 130–200 times sweeter than sucrose. It is stable at high temperatures, so it can be used to sweeten baked goods. In aqueous solutions and when stored under cool dry conditions, it has a virtually unlimited shelf life. Low-calorie soft drinks formulated with acesulfame K show no decrease in sweetness over a period of several months. Sterilization and pasteurization do not affect the taste of this sweetener.

Stevioside and thaumatin are natural sweeteners that are permitted in several countries, and neohesperidin dihydrochalcone (NDHC), alitame, sucralose and L-sugars are undergoing testing for regulatory approval. Approvals have been given or are expected for these additives.

Thickeners

Thickeners are among the most commonly used additives, as they exert an effect on the texture and viscosity of food and drink products. They are hydrocolloids that are soluble in water or can be readily hydrated or dispersed. Thickeners interact with water in food through their fibrous, or cross-linked, structure and their many polar groups, especially hydroxyl groups. Water molecules that have a polar character are oriented around the polar groups of the thickeners. Formation of a hydration layer of this kind, often accompanied by unfolding of the molecules, limits the mobility of the water so that the viscosity of the system increases. Thickeners are generally added under defined conditions to water or to a food component with high water content.

The main uses of thickeners in the food industry are in fermented milk products like yogurt, in the manufacture of ice cream and in fat emulsions. The thickeners commonly used in food are polysaccharides of plant, microbial, or semi-synthetic origin, e.g., milled seeds, extracts of sea algae and plant extracts or exudates. The most important thickeners commonly used in foods are alginic acid, alginates, agar, carob, guar, tragacanth, acacia and xanthan gums, pectin, cellulose esters, starch, starch esters and modified starches.

Processing Aids

Each foodstuff consists of chemicals that are more or less characteristic of it. Because of natural variations, however, it is frequently necessary to adjust the composition to provide a product of constant quality. Some food additives are used to aid the crystallization of sucrose or glucose, to modify starch and gluten in the production of stable dough and to promote protein and pectin precipitation. Some food additives are used as leavening agents, as carriers of other food ingredients, or as pH regulators, plasticizers, clarifying agents, clouding agents and drying agents. According to the CAC, processing aids are defined as:

> "Any substance or material, not including apparatus or utensils and not consumed as a food ingredient by itself, intentionally used in the processing of raw materials, food or its ingredients, to fulfill a certain technological purpose during treatment or processing and which may result in the non-intentional but unavoidable presence of residues or derivatives in the final product."

The Codex Committee on Food Additives has prepared an inventory of processing aids in which the functional effects, chemical names or descriptions, area of utilization, level of residues, interaction with foods and toxicological evaluations of processing aids are described.

9

CHEMOPREVENTERS IN THE DIET

Diet has been linked with cancer for centuries. In the Middle Ages, it was commonly believed that yeast preparations could prevent cancer. As early the 1500s, it was recommended that cancer patients should eat cucumber and pumpkin and avoid grilled or fried meat, fish in jelly and raw eggs.

In the 1980s, research began to more closely investigate the possible links between cancer and diet. A study conducted in the U.S. showed that the main factors that contributed to cancer deaths were diet, tobacco and infections, and that factors such as drugs, industrial products and food additives had relatively insignificant effects. Epidemiological studies suggest that diet may influence 30–40% of all cancers in men and 60% in women. It was also reported that the best-validated risk factor is high intake of fat, especially saturated fat, which correlates with an increased risk for colorectal cancer and the risk for prostate cancer and possibly ovarian, renal and pancreatic cancer. The newest information was provided by large-scale population studies and experiments on animals. As part of the Iowa Women's Health Study, more than 41,000 postmenopausal women completed an extensive food-frequency questionnaire (127 food items including 29 vegetables and 15 fruits) in 1986. During 4 years of followup, 138 women were diagnosed with lung cancer. These women were compared with 2814 other women randomly selected from the large study group. High intakes of all vegetables and fruits; all vegetables and green leafy vegetables; high intake of vitamin C vegetables and fruits, carrots and broccoli were associated with lower lung cancer risks. The associations were stronger in former smokers than in current smokers.

In a study conducted in Switzerland, 107 women with breast cancer and 318 control women admitted to the same hospitals for treatment of

Table 9.1   Mutagenic and Carcinogenic Substances Found in Foods

Origin	Examples
Plant	hydrazines, gossypol, pyrrolizidine alkaloids, safrole
Fungal	aflatoxin, patulin
Environmental	arsenic, dioxin, DDT, PCBs, radiation
Formed during processing	fatty acid hydroperoxides, nitrosamides, nitrosamines, PAHs

other diseases were interviewed with a questionnaire about their frequency of consumption of 50 indicator foods. The study showed that increased intake of total energy (calories) and frequent consumption of meat, cheese or alcohol were with increased risk of breast cancer. Significant protective effects, several specific, were seen for total green-vegetable consumption.

The role of diet in causing cancer can be summarized as follows:

- An elevated intake of dietary fat or calories
- Frequent use of roasted, smoked or salted products
- Consumption of food involving mutagenic or carcinogenic agents

The origins and examples of mutagenic or carcinogenic agents that may be present in foods are shown in Table 9.1.

Although it is well known that many dietary factors contribute to carcinogenesis, a negative relation between cancer risk and diet is also estimated in several studies and current scientific interest in recent years has been directed toward the cancer-preventing potential of naturally occurring constituents of the diet.

CHEMOPREVENTION

The food components that have genotoxic (antimutagenic and anticarcinogenic) potentials are called *chemopreventers* (CPs). Cancer chemoprevention can be defined as "prevention of cancer by the administration of one or more chemical entities, either as individual drugs or as naturally occurring constituents of the diet." The effects on cancer risk of adverse and beneficial substances in the diet are illustrated in Figure 9.1.

As can be observed from the figure, appropriate dietary measures can modulate cancer risk downward either by decreasing the amount of adverse substances or increasing the amount of beneficial substances taken into the body.

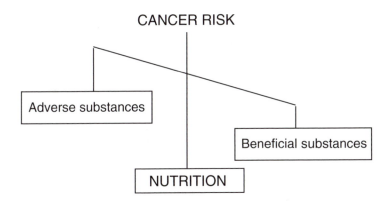

CANCER RISK

Adverse substances

Beneficial substances

NUTRITION

Figure 9.1   Balancing the diet to reduce cancer risk. (From Verhagen et al. 1993, with permission of Miller Freeman BV.)

The National Cancer Institute (NCI) in the U.S. has a division of cancer prevention with a chemopreventive department. NCI seeks to develop chemoprevention against the *initiation, promotion* and *progression* stages of carcinogenesis. Figure 9.2 shows the main stages that occur in development of carcinogenesis.

Cancer cells are the result of multiple genetic defects resulting from exposure to infections, environmental agents and dietary constituents. The initiation stage of cancer development can occur several thousand times in the body's cells, but, for a certain period of time, the cell-repair system repairs the damage until such time as the defense might fail to function. Anti-initiation substances are mostly enzymes already present in the body. These enzymes are biocatalysts that are able to transform carcinogenic substances into a water-soluble form and eliminate them from the body before they can cause any harm. Other enzymes help the amino acid gluthathione in its detoxifying activity. Chemical carcinogens are made water-soluble and excretable in two different forms of enzymatic reactions. The first type is hydroxylation, i.e., transport of a hydroxyl (OH) group into the potential carcinogenic molecule. The other type of reaction is a combination of the carcinogenic molecule with readily water-soluble molecules. Unfortunately, hydroxylation often leads to formation of epoxides, which are in themselves highly carcinogenic. Therefore, NCI seeks substances that would activate the second type of reaction, the detoxifying mechanisms of the body's own protective enzymes, such as gluthathione-S-transferase, which quenches the radicals.

In the promotion stage, a number of chain reactions take place that result in the cell's becoming a cancer cell. The normal defense mechanisms of the organism, which are responsible for cell division, lose control

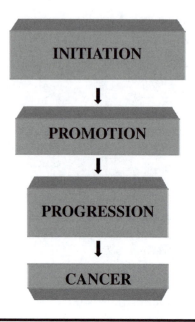

Figure 9.2   Stages in cancer development. (Adapted from Tolonen, 1990.)

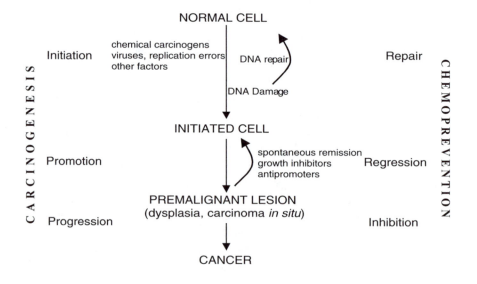

Figure 9.3   Multistep modeling of carcinogenesis. (From Singh, V.N. and Gaby, S.K., 1991, with permission of *Am. J. Clin. Nutr.*)

of this cell and the cancer cell begins to divide itself uncontrollably. A tumor begins to form genetically identical cells that are referred as "clone mass." Substances that cause cell multiplication are known as "promoters." Any substance that inhibits pathological cell division can, in principle, be considered an "anti-promoter." Examples of such substances are selenium, β-carotene and vitamin A (natural retinol as well as synthetic retinoids).

In the progression stage, the initiated cell population produces clones of cells that are slightly different genetically. These cells show a high division rate, as well as capacity to infiltrate into other tissues and to send metastases to other parts of the body. The cells appear to be individually changed and disorganized, and this situation is called *dysplasia*. Chemoprevention of the progression is based on the hypothesis that a second genetic damage is required for the promoted cell to turn into a fully developed cancer cell. Principally it would be possible to prevent this second change like the initiation in the first place.

METHODS FOR STUDYING ANTIGENOTOXICITY

Short-Term Tests

These tests can be conducted *in vitro* like a reverse mutation test with bacteria (Ames test) and a clastogenecity test with eukaryotic cell systems (chromosomal aberration test). The results of *in vitro* tests should be confirmed with *in vivo* tests conducted by using experimental animals. One of the most relevant biomarkers of genotoxicity and, potentially, carcinogenesis, is the occurrence of mutations.

Long-Term Tests

The anti-carcinogenic potential of a compound can be determined by lifetime exposure, such as in the diet, of experimental animals to various dose levels of the test compound up to the level of its chemopreventive effect as well as the level of its own toxic effects. Data indicate that carcinogens are highly specific with regard to their target tissue in inducing both tumors and mutations. This specificity may reflect the dependence on tissue-specific metabolic activation, the organ-specific environment or both. Ideally, therefore, mutation should be determined in a real animal rather than in a cell-culture system. In a recent study, it was found that the lacI transgenic rodent model provides such a system. This model is used to investigate tissue, species and sex specificity of mutation induced by selected dietary carcinogens and to examine how some compounds may alter the induction of mutation. It was also shown that some dietary compounds could modulate the mutagenic potency of these chemicals, thus, the lacI transgenic rodent is suggested as a useful model for the study of chemoprevention *in vivo*.

For antimutagenic and anticarcinogenic substances, the beneficial potential (LBEL) and toxicological endpoints (NOAEL) should be considered in a single evaluation because such a beneficial effect is valuable only in the absence of toxicity. It should be confirmed that the beneficial effects of the chemopreventers should be evident at (much) lower dose levels than where its own toxicity as a chemical is expected.

Epidemiological Studies

These studies are based on questionnaires dealing with food intake and dietary habits of the humans being studied. Negative associations with cancer incidence have been reported with fiber, fruit and vegetable consumption and caloric restriction.

Biomarker Studies

Biomarker studies have gained importance in studying antimutagenic and anticarcinogenic effects of chemical substances. In a study conducted at TNO Nutrition and Research Institute in Netherlands, ten healthy male volunteers, all nonsmokers, were asked to refrain from any foods containing glucosinolates and consumed standardized dinners containing 300 g of glucosinolate-free vegetables for a control period of 3 weeks. During a subsequent 3-week intervention period, five volunteers continued the glucosinolate-free daily dinner (control group), while the other five (sprouts group) consumed 300 g of Brussels sprouts. On two consecutive days in the midst of the control period and the intervention period (days 13/14, 34/35 of the study) blood samples were drawn for the quantitative determination of detoxifying gluthathione-S-transferase isozymes. Results indicate a ca. twofold increase in the levels of alpha-class glutathione-S-transferases in the sprouts group during the intervention period.

MECHANISM OF ACTION OF CHEMOPREVENTERS

Although the numbers of chemopreventers (CPs) increase every day, knowledge about their effect mechanisms remains limited due to the complexity of their activities. The described effects may be the result of a single reaction as well as the results of subsequent reactions. Table 9.2 examines the mechanism of effects of CPs in two major categories. These effects, alone or in combination, reduce the hazardous effects of mutagens and carcinogens in the body. In most of the studies conducted on this subject, the antioxidative characteristics of CPs seem to play the most significant part in their protective activity. Their effect against different cancers may not be equally beneficial because they exhibit different mechanisms of action.

As can be seen in Table 9.2, some CPs can act extracellularly by reducing the formation and bioavailability of carcinogenic species, while some act intracellularly by influencing particular enzyme systems. In other studies, the postulated mechanism for inhibiting the action of carcinogens is described in Table 9.3.

Table 9.2   Mechanism of Action of Chemopreventers

Extracellularly	*Intracellularly*
During the preparation of foods ■ Inhibiting the formation of M/C[a] Effects in the intestine ■ Formation of non-M/C complexes ■ Reducing bioavailability ■ Diluting with dietary fibers ■ Increasing adsorption on other food components ■ Accelerating intestinal transit ■ Protecting the mucosal barrier ■ Modifying intestinal microbial flora ■ Inhibiting the penetration of cells by M/C	At cellular level ■ Enhancing the activities of enzymes involved in detoxification of M/C ■ Inhibiting the activities of enzymes involved in formation of M/C metabolites ■ Trapping of electrophils ■ Scavenging reactive oxygen species ■ Inhibiting metabolic activation ■ Protecting nucleophilic sites of DNA ■ Inhibiting the detrimental effect of procarcinogens on DNA

[a] M/C = mutagens and/or carcinogens.

From Stavric, 1994, with permission from Elsevier Science.

As observed in Table 9.3, the cellular activity of carcinogens can be inhibited in a variety of ways, including the inhibition of carcinogen formation as with N-nitroso compounds, and the inhibition of activation for those carcinogens requiring metabolic activation.

Some nutrients with chemopreventive potentials, their proposed modes of action and major food sources are shown in Table 9.4, where it can be observed that vitamins A, C, E, β-carotene and selenium have anticarcinogenic activity due to their antioxidative properties.

Table 9.3 Postulated Mechanism for Inhibiting the Actions of Carcinogens

I. Inhibition of carcinogen formation from precursors (nitrosoformation from reaction of nitrate with amine or amide)

Examples: ascorbic acid, tocopherols

II. Blocking agents: Inhibition of carcinogen activation, induction of detoxification or limiting carcinogen reaction with critical cellular molecules

■ Inhibition of carcinogen activation
Examples: disulfiram, 7,8-benzoflavone

■ Induction of detoxifying enzymes
Examples: butylated hydroxyanisole, phenols, coumarins, flavones

■ Enhancers of glutathione-S-transferase activity
Examples: coumarin, benzylisothiocyanate

■ Trapping of reactive carcinogenic electrophiles
Examples: glutathione, methionine

III. Inhibitors acting subsequent to carcinogen exposure-suppressors

Examples: retinoids, plant sterols, protease inhibitors

From Carr, 1985, with permission.

SIGNIFICANT CHEMOPREVENTERS IN THE HUMAN DIET

Antioxidant Vitamins

Vitamin A and Carotenoids

In recent years, considerable attention has been paid to the possible inverse link between vitamin A (retinol) and the occurrence of certain forms of cancer. In laboratory animals, it has been determined that vitamin A or the retinoids can prevent cancer of the skin, lung, urinary bladder, breast, esophagus and stomach. In a 10-year prospective study of more than 250,000 Japanese adults, an association between frequency of green and yellow vegetable consumption and rates of mortality from cancer at different sites was found. In Japan, green and yellow vegetable consumption accounts for an average 44% and 23% of the typical dietary intake of vitamins A and C, respectively. This study and others suggested that vitamin A might be exerting a protective role.

After the reports indicating that vitamin A might be a potent anticarcinogen, interest began switching to the idea that the fruits and vegetables

Table 9.4   Postulated Nutritive Dietary Anticarcinogens, Their Proposed Mode of Action and in What Foods They Are Found

Anticarcinogen	Proposed Mode of Action	Major Food Sources
Vitamin A	cell differentiation	Margarine, dairy products
β-carotene	antioxidant, provitamin A	Yellow-green vegetables
Vitamin C	antioxidant	(citrus) fruits, vegetables
Vitamin E	antioxidant	vegetable oils, whole meal
Selenium	antioxidant	meat and meat products, eggs, dairy products
Calcium	binding of bile acids and fatty acids	dairy products

From Verhagen et al., 1993, with permission of Miller Freeman BV.

containing carotenoids, which are precursors of vitamin A, might be a defense against cancer. *Carotenoids* are free-radical traps and remarkably efficient quenchers of singlet oxygen, which is mutagenic and particularly effective at causing lipid peroxidation. Carotenoids have been shown to be anticarcinogenic in rats and mice. β-*carotene, canthaxanthin, crocetin* inhibited UV-induced skin cancers in mice when given orally. It was also demonstrated that β-carotene and canthaxanthin inhibited 1-methyl-3-nitro-1-nitrosoguanidine (MNNG)-induced stomach cancers when given orally to mice and rats. Experiments on rats have shown that β-carotene may also inhibit colon and pancreatic cancers. β-carotene is present in carrots and in all foods that contain chlorophyll. A reduction of approximately 50% in the risk of lung cancer was observed to be associated with high consumption compared with low intake of carotene-containing fruits and vegetables. Regular dietary intake of vitamins C and A was found to be protective for both colon and rectal cancer. In a study conducted in the U.S., blood samples provided from more than 25,000 adults were frozen for later analysis. During 15 years of follow-up, just 28 individuals were diagnosed with oral or pharyngeal cancer. The study showed that the carotenoid levels were consistently lower in cases than in controls. By reacting with free radicals, carotenoids should theoretically prevent the damaging effects of these free radicals expressed in terms of lipid peroxidation, enzyme destruction and damage to both RNA and DNA. It is the latter damage, in terms of altering the genotype of cells, that is supposed to link excess free-radical production with cancer. *Lycopene*, which is a carotenoid that acts as the coloring pigment in tomatoes, was shown to be an even more potent inhibitor of endometrial, lung and mammary cancer cell growth than β-carotene in *in vitro* and *in vivo* studies.

Vitamin C (Ascorbic Acid)

Animal studies with regard to prevention showed that the reduced form of ascorbic acid is effective in blocking the formation of active forms of chemical carcinogens, or in blocking the carcinogenesis due to free-radical generation caused by radiation or by hormonal or viral carcinogens. A study on RIII mice, a strain in which the incidence of mammary tumors is 56%, showed that spontaneous mammary tumors were not completely prevented by ascorbic acid; however, they were significantly delayed and reduced in incidence, increasing median tumor-free life span by 50%. Two groups of studies have shown that ascorbic acid or its lipophilic derivative, ascorbyl palmitate, was effective in blocking the skin tumor promotion.

Ascorbic acid, as well as retinyl palmitate, significantly inhibited the development of neoplastic pulmonary lesions in mice exposed to fiberglass dust. Other carcinogens such as anthracene, benzopyrene, heavy metals, organochlorine pesticides-DDT, dieldrin and lindane, have been blocked by ascorbic acid.

Epidemiological studies suggest that vitamin C is protective against several types of cancer in humans. Vitamin C, which is also an antioxidant, acts effectively as a scavenger of nitrite in the stomach, thereby preventing the formation of carcinogenic nitrosamine species and of free radicals formed during the processing of foods or in metabolic processes. It was shown to be anticarcinogenic in rodents treated with ultraviolet radiation, benzo(a)pyrene and nitrite. In laboratory studies, vitamin C has been shown to inhibit formation of carcinogenic nitroso compounds, to prevent malignant transformation of cells grown in culture and to lead to regression of transformed cells. In humans, several studies have found that consumption of foods containing appreciable amounts of vitamin C is associated with a lower risk of esophagus, pancreas and stomach cancers as well as those of the oral cavity and pharynx. In a case-control study of stomach cancer performed in Sweden, 338 stomach cancer patients and 679 controls (40–79 years of age) living in the same areas were interviewed, using a food frequency questionnaire that enquired about diet during two periods of life — adolescence and the 20 years leading up to the interview. Dietary habits of the past 20 years were of particular interest, because the induction time of stomach cancer is estimated to be about 20 years. It was observed that dietary intakes of vitamin C and β-carotene had preventive potential in stomach cancer risk; the effects were especially strong for intakes of these vitamins 20 years prior to interview. However, in multivariate analysis, only vitamin C retained a significant protective effect against stomach cancer.

In studies for which an index of vitamin C was reported or plasma concentration determined, the findings strongly support that a protective effect is likely for cancers of the larynx, lung, rectum, cervix and breast and such a role for vitamin C is possible for the cancers of the colon and bladder.

Vitamin E

Vitamin E refers to a group of eight fat-soluble compounds, tocopherols and tocotriends, of which α-tocopherol is the biologically most active form. The richest sources are vegetable oils (wheat germ, soybean, corn), margarine, nuts, seeds, cereal grains (especially wheat germ) and vegetables. It has been hypothesized that free-radical-initiated cell damage, particularly DNA damage, is involved in the etiology of cancer and that vitamin E, because of its property as an antioxidant and free-radical scavenger, may play a role in the prevention of cancer. Animal and *in vitro* studies suggest that vitamin E decreases tissue susceptibility to malignancies in several stages of carcinogenesis by a number of ubiquitous mechanisms. These include inhibition or blockage of formation of nitrosamines or other mutagens or carcinogens from precursors via direct chemical interaction. It has also been suggested that vitamin E inhibits mutagens or carcinogens from reaching or reacting with target sites by scavenging mutagens or by enhancing detoxification processes. Furthermore, the vitamin may prevent cancer by enhancing the immune response to developing cancer cells and may cause tumor regression by direct action on the cancer cells.

In an animal study conducted on the hamster buccal pouch carcinoma model for the study of solid malignant tumors, which have almost total similarity to the human counterpart both in terms of development and biological behavior, when a low-dose DMBA regimen was applied in the course of an extended process of carcinogenesis, vitamin E was able to completely prevent the development of epidermoid carcinomas. Vitamin E was also shown to cause regression of established carcinomas of hamster buccal pouch when injected alone into the tumor, or when administered orally in combination with β–carotene. In most animal studies, vitamin E has been shown to provide protection against oral and skin cancer, but against breast cancer only in certain circumstances. Some studies showed that the vitamin might be more effective given topically or by repeated oral dosing than via the diet. Combining with PUFA, selenium and vitamin C increases the effect of vitamin E. It has been demonstrated in animal experiments and in cell cultures that vitamin E has anticarcinogenic properties. Vitamin E is the major radical trap in lipid membranes and inhibits the carcinogenicity of the quinones adriamycin and daunomycin. The protective effect of vitamin E against breast and some gastrointestinal cancers has been observed in several epidemiological studies. It reduces free-radical formation, which, in turn, may result in a decrease of carcinogenesis. The anticarcinogenic potential of vitamin E may also be due to its inhibitory effects on proteases. Additionally, vitamin E inhibits both *in vitro* and *in vivo* formation of nitrosamines. In a study conducted on a total of 290 male oral cancer patients and 133 male esophageal cancer patients, vitamin E supplementation appeared to exert a protective effect among oral cancer cases. There was a synergistic protective effect of vitamin E and vegetable consumption in these cases. A significant inter-action between serum vitamin E and serum selenium levels with respect to hormone-related cancers, in particular to breast cancer, was detected. Subjects with low serum levels had a significant, 10-fold higher risk of breast cancer. This risk was also observed for the cancer of the upper gastrointestinal tract.

Allium Vegetables

Allium vegetables, such as garlic and onions, are traditionally known for their antibacterial and fungicidal properties. Their beneficial effects are related to specific sulfur-containing compounds that are responsible for the typical odor and flavor of these plants. *In vitro* models have also shown that some of these organosulfur compounds have antimutagenic activity and *in vivo* test systems have demonstrated anticarcinogenic activity. Protective effects of *allium* vegetables such as garlic, onions and leek are reported in some studies. Results from case-control studies and laboratory tests indicate that consumption of *allium* vegetables may considerably reduce the risk of stomach cancer. The results of a large-scale cohort study on diet and cancer performed in Netherlands studied the association between onion and leek consumption, garlic supplement use and the incidence of stomach carcinoma. Data have revealed a strong inverse association between onion (*Allium stellatum*) consumption and carcinoma in all parts of the stomach, except in the cardia. Garlic (*Allium sativum*) was shown to promote inhibition of nitrite-reducing bacteria and reduce nitrite formation, and was suggested as a protective factor against gastric cancer. A study conducted in China showed that, with the introduction of 10 g of fresh garlic (in homogenate form), the rate of reduction (90%) of nitrite concentration in gastric juices was significantly higher than in controls. Epidemiological studies done in China have shown that garlic components, especially allicin (allyl 2-propenethiosulfinate), inhibit the growth of transplantable tumors and reduce the incidence of certain spontaneously occurring tumors. In a dose–effect study where the *in vitro* cytotoxic effects of fresh garlic extract and diallyl trisulfide on two human gastric center cell lines was investigated, fresh garlic and diallyl trisulfide (allicin analog) were found to be more potent than 5-fluoruracil or mitomycin C in killing cancer cells. Components of garlic have also been found to inhibit the activity of diverse chemical carcinogens during both the initiation and promotion phases of carcinogenesis. The antimutagenic effects of garlic compounds are also determined by *in vitro* and *in vivo* experiments. Among these compounds, diallysulfide and diallyldisulfide were found to have antigenotoxic potential *in vivo*.

Dietary Fiber

Dietary fiber includes a number of polysaccharides and lignin that are not digested by the enzymes of the human gastrointestinal tract. These substances have specific chemical structures with varying physical and chemical properties:

- *Cellulose*, the most abundant molecule in nature, is also the principal structure of cell walls. Insoluble in water, it is found abundantly in the bran of cereal grains.
- *Lignin* is the water-insoluble component that makes up the woody part of plants. Legumes and fruits with seeds and the lignified cells of pears are important sources of lignin.
- *Hemicelluloses*, found in the cell walls of many plants, are soluble in hot water and occur in a variety of plant foods such as carrots, cabbage, celery, leafy vegetables, apples, melons, peaches, pears and whole-grain cereals.
- *Pectin*, which is soluble in hot water, has the capacity to hold water and form gels. Apples and citrus fruits are rich in pectin.
- *Mucilages and gums*, the nonstructural substances of plant cells, are soluble in hot water. In a study where *in vitro* binding capacities of various dietary fibers against indirect mutagens 2-amino-3-methyl-3H—imidazo (4,5-f) quinoline (IQ) were examined, it was found that potato fiber showed the highest binding capacity followed by pectic acid, glucomannan and cellulose.

The components of crude fiber are postulated to exert a protective effect against colon cancer by several mechanisms: (1) shortening intestinal transit time, thereby reducing the exposure time of epithelial surfaces to potential carcinogens; (2) influencing bile acid metabolism, resulting in decreased formation or enhanced excretion of potential carcinogens; (3) influencing intestinal flora with decreased degradation of bile acids and neutral sterols and (4) diluting potential carcinogens in the bowel. The meta-analysis of epidemiological data from 23 case-control studies strongly supported the findings that a 40% reduction in risk of colon cancer occurred among individuals consuming diets with high vegetable and grain content. Similar conclusions were drawn from another case-control study (1150 cases), in which it was found that the combination of high dietary fiber intake from any source (fruit, vegetables, cereals) or a high vegetable intake (especially, the crucifers) was uniformly protective for both colon and rectal cancers. Substantial protection against breast cancer was observed when a dietary pattern that combines low intake of fat with high intake of fiber and fermented milk products was used.

Flavonoids

Flavonoids are diphenylpropanes ($C_6C_3C_6$) that occur in edible plants and are a common component in the human diet. Flavonoids consist mainly of anthocyanidins, flavonols, flavones, catechins and flavanones. In addition to carotenoids, they are the major sources of red, blue and yellow pigments in plants. In a large number of epidemiological studies investigating the relationships between diet and cancer, a protective effect of the consumption of vegetables and fruits against various forms of cancer is generally attributed to vitamin C and β-carotene. However, the significance of flavonoids in vegetables and fruits is also important. These substances show antioxidant activity by scavenging the lipid peroxyl radicals, binding metal ions and inhibiting enzymatic systems responsible for free-radical production. A study conducted on flavone, which is a naturally occurring flavonoid distributed particularly in edible and medicinal plants, has shown that this substance is an inhibitor of the formation of IQ-type mutagens in glycine/creatine/glucose-refluxing model systems. The inhibitory effect of flavone in the heated model system was attributable to the reduction of the formation of Maillard reaction products. These findings could be a useful means for reducing the formation of mutagens of the 2-amino-3-methyl-3H-imidazo(4,5,f) quinoline (IQ) type during food processing.

It is commonly accepted that citrus fruits and juices are health-promoting foods. Among the citrus flavonoids, the polyhydroxylated flavonoids such as quercetin inhibited carcinogenesis in a number of models and selectively inhibited a variety of tumor cells' growth. The effects of dietary supplementation with the antioxidants ellagic acid, quercetin and vanillin were examined using a medium term multi-organ carcinogenesis model in rats. The results indicated that, while ellagic acid and quercetin exerted potent chemopreventive action in both the initiation and promotion stages in the present experimental system, their beneficial effects were restricted to the small intestine. Because small intestinal carcinomas are very infrequent in humans, the advantages of these phenolic compounds for human application as chemopreventors should not be overestimated. Quercetin is also found in onions, apples, tomato, beans, kale, broccoli, olive oil, red wine and tea. Besides quercetin, other food-derived flavonols such as kaempferol and myricetin have also been shown to have *in vitro* and *in vivo* antimutagenic and anticarcinogenic effects. Kaempferol is found in kale, endive, leek and turnip greens, whereas myricetin is found in French broad beans. Polymethoxylated citrus flavonoids such as tangeretin and nobiletin are shown to be more potent inhibitors of tumor cell growth than hydroxylated flavonoids. This difference in activity is suggested to be due to greater membrane uptake of the polymethoxylated flavonoids because methoxylation of the phenolic groups decreases the hydrophilicity of the flavonoids.

Isoflavonoids and Lignans

The importance of isoflavonoids in cancer prevention was noticed when it was observed that the incidence of breast cancer in women of the Pacific basin was 5–8 times lower than in the U.S., although no genetic differences in susceptibility were present between the two populations. It was suggested that the large-scale consumption of soybean could have a role in the lower incidence of breast cancer. In animal experiments, it was observed that chemically induced mammary tumors could be decreased by administration of soybean in the diet. It is suggested that the isoflavonoids occurring in high amounts in soybeans and soy products such as tofu, soy milk and miso; the lignans derived from whole grain bread; various seeds like linseed and sesame seed; fruits, berries and vegetables affect the intracellular enzymes, protein synthesis, growth factor action, malignant cell proliferation and angiogenesis. The anticarcinogenic activities of these compounds have been demonstrated in many *in vitro* cell cultures and *in vivo* animal studies. The isoflavonoid genistein, which has antioxidative properties, was shown to inhibit proliferation of numerous different types of malignant cells in culture. It was also found that genistein may act *in vivo* by blocking additional stages of breast cancer progression such as those stages resulting in invasion and metastasis. Epidemiological evidence obtained in Japan has shown lower colon cancer incidence in areas with high tofu consumption. In a recent study of the effect of diets containing soy products in inhibition of the early stages of azoxymethane-induced colon cancer in F344 rats, it was found that 0.015% genistein was more effective than soy flour and full-fat soy flake diets containing 0.049% genistein derivatives (primarily glycosides) in inhibiting the formation of precancerous lesions assessed as foci with aberrant crypts (FAC). It was concluded that eating soybeans and soy flour might reduce the early stages of colon cancer. *In vitro* studies with prostate cancer cells in culture have also demonstrated that genistein and daidzein inhibit cell proliferation. Using the developmentally estrogenized mouse model, it has been shown that the development of dysplastic changes in prostate was delayed by soy feeding.

Lignans are diphenolic compounds that contain dibenzylbutane skeleton. Lignans are similar in structure to estrogens, isoflavonoid phytoestrogens and tamoxifen, which has weak estrogenic/antiestrogenic properties. Tamoxifen competes with estradiol in binding to nuclear type II estrogen-binding sites of breast tissue, thus inhibiting estradiol-stimulated growth of cancer cells. However, it acts like a weak estrogen on other body systems (other than breast tissue). It is even thought that women who take tamoxifen as a drug may share some of the beneficial effects of estrogen-replacement therapy, such as a decreased risk of osteoporosis and heart disease. Flaxseed contains 75 to 800 times more lignans than more than 60 other plant foods, making it the richest known source of lignans. Lignans are also present in legumes, cereal brans, cereals, vegetables and fruits. Several observations have indicated that lignans might protect against cancer. Some epidemiological studies showed that the urinary excretions of lignans and equol was lower in breast cancer patients than in vegetarians, and very high excretion of lignans and equol were detected in chimpanzees, animals that are highly resistant to breast cancer. Other studies have also shown that breast cancer patients and individuals at high risk for breast and colon cancers excrete significantly lower levels of mammalian lignans than vegetarians or those at lower risk of developing these diseases. High lignan excretion was also detected in human subjects living in the areas of the world with lower colon cancer risk. Both breast and colon carcinogenesis was reduced in rats fed flax seed containing high amounts of secoisolariciresinol.

A recent study investigated the ability of the isoflavonoids genistein and equol; the lignans enterodiol, enterolactone, nordihydroguaiaretic acid (NDGA); and the lignan metabolite methyl p-hydroxyphenyllactate to interfere with mitogenic and tumor promotional signal transduction pathways. The results suggest that both genistein and equol interfere similarly with phorbol ester 12-O-tetradecanoylphorbol-13-acetate TPA-induced signal transduction pathways. If the ability to antagonize phorbol ester effects is important for chemopreventive efficacy, then equol and genistein might possibly be equally effective CPs.

Polyphenolic Substances

It was reported in several studies that *ellagic acid*, a natural polyphenol abundant in many vegetables, in fruits such as grapes and strawberries, and in nuts and other foods, could be effective in preventing the development of cancer induced by tobacco carcinogens. Ellagic acid and another plant polyphenol, *chlorogenic acid,* a normal component of coffee beans, instant coffee, blueberries and peaches, were found in several animal studies to be potential CP agents against several carcinogens. The major beneficial component in green tea has been identified as polyphenol *epigallocatechin-3-gallate (EGCG)*, a compound that possesses antioxidant properties. This substance is a common constituent of Japanese green tea, but appears only in small quantities in black tea. The EGCG (or green tea extract) has been shown in laboratory animal studies to reduce different types of spontaneous or chemically induced tumors such as tumors of the liver, stomach, skin, lungs and esophagus.

Catechin, a flavan-type polyphenol, found in sorghum grain and fava beans, was found to be both an antimutagen and an anticarcinogen.

Eugenol, the main component of oil of cloves, is also present in the essential oils of many other plants such as cinnamon, basil and nutmeg; *trans-anethole* is the main component of the volatile oils of fennel and Chinese star anise. In a study where the effect of liver S9 from eugenol-treated rats on established mutagens in the Ames test was investigated, it was found that the mutagenicity of benzo(a)pyrene was lower when using liver S9 fractions prepared from rats treated with eugenol (1000mg/kg b.w.) than using liver S9 from control rats. In another study where the effects of trans-anethole and eugenol on drug-metabolizing enzyme activities in rat liver was observed, it was detected that both eugenol and trans-anethole induced phase-II biotransformation enzymes, a characteristic of type-A inhibitors of carcinogenesis.

Selenium

Selenium (Se), an essential element, is found in amounts that vary widely in different soils and, thus, in foods and feeds produced on these soils. Although plants and forages consumed by animals may accumulate levels of selenium in excess of 1000 ppm, normal human diets rarely exceed the recommended safe level, which is 3–4 ppm. Selenium is found in grains, meat, seafood and certain vegetables. The chemopreventive action of selenium is probably attributable to its antioxidant properties and its involvement with the enzyme gluthathione peroxidase, an enzyme essential for preventing oxygen radical-induced lipid peroxidation. Another possible anticarcinogenic mechanism of Se is the inhibition of cell proliferation. Several studies have found that microgram quantities of dietary selenium can effectively reduce skin, colon, liver and mammary tumors in experimental animals. Because selenium content of an individual food can vary widely due to geographic variations, Se levels in hair, blood, nails or urine are used as biomarkers in human studies. In a prospective cohort study on diet and cancer that started in 1986 on 120,852 Dutch men and women aged 55–65 years, toenail Se contents were used as long-term markers to evaluate the association between Se status and lung, gastrointestinal and breast cancers. After 3.3 years of follow-up, the study showed that Se had protective effects against lung and stomach cancers. The levels of Se in toenails were not associated with the risk of breast, colon or rectum cancers and it seems that a protective potential of selenium, if any, is limited to epithelial cancers. The results of randomized trials of Se and primary liver cancer conducted in China also provided evidence that Se may prevent cancer at this site. Because Se may have toxic effects at levels only to 4 to 5 times the level normally ingested in the diet, very careful attention to dosage is necessary to avoid adverse effects.

Miscellaneous Chemopreventers

In recent studies, a possible role for increased *dietary calcium* and *vitamin D* in reducing risk of colonic and mammary cancers, even in the presence of high-fat diets, are evaluated. The studies have shown that dietary calcium and vitamin D in a high-fat diet induced adverse changes in mammary glands and several other organs that were reversed by increasing dietary calcium and vitamin D. It was also found in cell culture systems that exposure to calcium or vitamin D reduced the oncogenic properties of colon cancer cells. Administration of calcium has also been shown to reduce the incidence of colonic polyps that may affect the formation of colon cancer.

The modifying effects of dietary administration of *capsaicin*, which is the principal pungent capsicum fruit on azoxymethane (AOM)-induced colon tumorigenesis, were investigated in male F344 rats. In an aberrant crypt foci (ACF) bioassay, feeding of capsaicin at a dose of 500 ppm for 4 weeks significantly inhibited ACF formation induced by AOM (20 mg/kg body weight, once a week for 2 weeks). In a subsequent long-term study designed to confirm the protective effects of this compound on ACF development, dietary exposure of capsaicin during the initiation phase was found to significantly reduce the incidence of colonic adenocarcinoma, indicating that this substance is suggested to be useful for the prevention of human colon cancers.

Chlorophylline, the sodium and copper salts of chlorophyll, have also been shown to have antimutagenic activity. Because this substance is regarded as an antioxidant, its antimutagenic activity is suggested as a scavenger of free radicals or it may act by cooperating with the active part of mutagenic compounds. The antimutagenic activity of certain vegetable extracts has been correlated with their chlorophylline content. In a study where cancer prevention by dietary chlorophylline was investigated in a rainbow trout multi-organ tumor model, it was found that this substance is a reproducibly effective chemopreventive agent for dibenzo(a,l)pyrene multi-organ tumorigenesis in trout and it was suggested that reduced dibenzo(a,l)pyrene-DNA adducts may be predictive biomarkers of chlorophylline reduction of dibenzo(a,l)pyrene-initiated hepatic tumors.

Conjugated linoleic acid, which is an abundant compound in meat, dairy products and especially cheese, was shown to inhibit stomach and skin cancers in rats and cancer in the mammary glands of mice. The anticarcinogenic effect of this substance is suggested to be the result of its antioxidative activity.

Curcumin has been widely used as a spice and coloring agent in foods. Recently, curcumin was found to possess chemopreventive effects against skin, stomach, colon and oral cancers in mice. It was also found that curcumin inhibits N-diethylnitrosamine (DEN)-induced hepatocarcinogenesis in the mouse. Curcumin has a potent preventive activity during the diethylstilboestrol (DES)-dependent promotion stage of radiation-induced mammary tumorigenesis. Curcumin was also found to be an effective agent for chemoprevention, acting at the radiation-induced initiation stage of mammary tumorigenesis.

The GST-inducing and anticarcinogenic properties of cruciferous vegetables have been attributed mainly to the degradation products of *glucosinolates* and also to the presence of *dithiolethiones*. It is suggested that the anticarcinogenic effects of these substances may occur because of their ability to induce both phase I and phase II detoxification enzymes, presumably by interaction with the *Ah*-receptor at high dietary levels. In a recent study, the apparently specific reducing effect of Brussels sprouts on oxidative DNA damage and an induction of the detoxification enzyme glutathione S-transferases-α were determined.

Myristicin, which is a major chemical constituent of parsley leaf oil, was found to have the ability to induce an increase in the detoxifying enzyme glutathione S-transferase (GST)'s activity and this substance was suggested to be used as a potential chemopreventive agent. Induction of GST is usually assumed to result in a decreased cancer risk, especially in the liver and small intestine.

Phytic acid (myo-inositol hexaphosphate), which is a constituent of plants such as edible legumes, cereals, oil seeds and nuts, is suggested as one of the most promising cancer chemopreventive agents. In a study where the mechanism by which phytic acid expresses preventive action to cancer was investigated, it was found that phytic acid inhibited the formation of 8-oxo-7, 8-dihydro-2′-deoxyguanosine in cultured cells treated with an H_2O_2-generating system, although it did not scavenge H_2O_2. Phytic acid alone did not cause DNA damage and thus, it should not act as a prooxidant. It was concluded that phytic acid acts as an antioxidant to inhibit the generation of reactive oxygen species from H_2.

Phytic acid has also been shown to decrease the incidence of colon cancer in rats.

In a study where the antimutagenic activities of some foods against the direct-acting mutagen sodium azide have been investigated by the Ames test, it was found that *Urtica* sp. (nettle) and rosehip showed significant antimutagenic effects. The antimutagenic effect of *Urtica* sp., which is an herb used in cancer treatments, was suggested to be due to lignin-like compounds, fiber, polyphenols or other compounds that are heat-stable antimutagenic substances in vegetables and fruits. The antimutagenic activity of rosehip tea, which has been used in the treatment of hemorrhoids, eczema, fever and diarrhea, was suggested to be attributable to flavonoids such as tannins, catechins and proanthocyanidins, in which the Rosaceae family is rich.

SCOPE OF FUTURE RESEARCH ON CHEMOPREVENTION

As can be observed from the literature mentioned throughout this chapter, eating a low-fat, high-fiber diet and plenty of fruits and vegetables seems like the best approach for prevention of cancer. However, the data is not yet sufficient to express the quantity of the CPs that should be taken into the diet for actual prevention of carcinogenesis. The answers to such questions such as, "How many grams of fiber does it take to prevent colon cancer?" or "How many grams of fresh garlic will inhibit tumor formation in different parts of the body?" or "How many grams of selenium can prevent mammary tumors in women?" are not still clear. A wide range of studies involving many procedures is still required (Figure 9.4).

Such evaluations may produce a list of CPs in amounts that are safe in the diet. This list may involve single components as well as multiple dietary constituents and functional foods. In focusing diet, nutrition and cancer prevention, the effects of antimutagenic and anticarcinogenic substances in the diet should be considered in accordance with the changing dietary patterns and eating behaviors of different individuals living in different parts of the world.

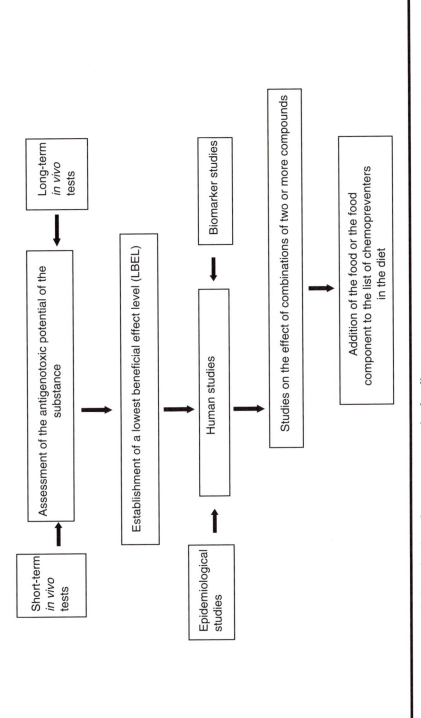

Figure 9.4 Framework for selecting chemopreventers in the diet.

REFERENCES

Abou-Donia, M.B., 1995. *CRC Handbook of Toxicology* (Derelanko, M.J. and Hollinger, M.A., Eds.) CRC. Boca Raton. 530-589.

Adlercreutz, H., 1996. *Natural Antioxidants and Food Quality in Atherosclerosis and Cancer Prevention* (Kumpulainen, J.T., Salonen, J.T., Eds.). The Royal Society of Chemistry. Cambridge, UK. 349-355.

Akagi K, Hirose M, Hoshiya, T., Mizoguchi, Y., Ito, N. and Shirai, T., 1995. Modulating effects of ellagic acid, vanillin and quercetin in a rat medium term multi-organ carcinogenesis model. *Cancer Lett.* 20; 94(1): 113-121.

Altuğ, T., Yousef, A.E. and Marth, E.H., 1990. Degradation of aflatoxin B_1 in dried figs by sodium bisulfite with or without heat, ultraviolet energy or hydrogen peroxide. *J. Food Protec.*, Vol. 53, 7, 581-583.

Ames, B.N., 1983. Dietary carcinogens and anticarcinogens — oxygen radicals and degenerative diseases. *Science*, 221: 1256-1264.

Anonymous, 1993a Serum micronutrients and the subsequent risk of oral and pharyngeal cancer. *AOV Newsletter* 6: 8-9.

Anonymous, 1993b. Vitamin supplement, use and risk for oral and esophageal cancer. *AOV Newsletter* 5: 8-10.

Anonymous, 1994a. Diet and breast cancer. *AOV Newsletter*. 8:9.

Anonymous, 1994b. Nutrients and gastric cancer risk. *AOV Newsletter*, 10: 6-7.

Anonymous, 1997a. Food consumption and exposure assessment of chemicals. Report of an FAO/WHO Consultation. WHO/FSF/FOS/97 5. Geneva. 1-6.

Anonymous, 1997b. Monograph on residues and contaminants in milk and milk products. International Dairy Federation Special Issue 9701. IDF General Secretariat. Brussels. 98-106.

Anselme, J., 1979. *N-Nitrosamines* (Anselme, J., Ed.). American Chemical Society. Washington, D.C. 1-12.

Ariéns, E.J. and Simons, A.M., 1982. *Nutritional Toxicology* (Hatchcock, J.N., Ed.). Vol. 1. Academic Press. New York. 17-80.

Aritsuka, T., Tanaka, K. and Kiriyama, S., 1989. Protective effects of beet dietary fiber against 1,2-dimethylhydrazine-induced carcinogenesis in rats. *J. Agr. Chem. Soc. Japan*, 63: 7, 1221-1229.

Aurand, L.W., Woods, A.E. and Wells, M.R., 1987. *Food Composition and Analysis*. Van Nostrand Reinhold. New York. 466-471.

Bauer, K., Garbe, D. and Surburg, H., 1988. *Ullmann's Encyclopedia of Industrial Chemistry* (Gerhartz, N., Ed.). 5th ed. Vol. A.11. Weinheim, FRG. 144-146.

Belinsky, S.A., Kauffman, F.C. and Thurman, R.G., 1987. *Nutritional Toxicology* Vol. 2, (Hathcock, J.N., Ed.). Academic New York. 41-42.

Bender, A.E., 1973. *Nutrition and Dietetic Foods*. Chemical Publishing Co., Inc. New York. 49-59, 87-96.

Berge, P., 1993. The EC flavor directive and the worldwide harmonization of flavor regulations. *Flavor and Fragrance J.*, Vol. 8, 81-85.

Betz, J.M., 1995. Plant toxins. General referee reports: *J. AOAC Int.*, Vol. 78, No: 1, 141-143.

Beuchat, L.R., 1998. Surface decontamination of fruits and vegetables eaten raw: A review. WHO/FSF/FOS/98.2. WHO. 11-12.

Biles, J.E., McNeal, T.P. and Begley, T.H., 1997. Determination of Bisphenol A migrating from epoxy can coatings to infant formula liquid concentrates. *J. Agric. Food Chem.* 45, 4697-4700.

Block, G. and Schwarz, R., 1994. *Natural Antioxidants in Human Health and Disease* (Frei, B., Ed.). Academic Press Inc. San Diego. 129-155.

Boer, J.G., 2001. Protection by dietary compounds against mutation in a transgenic rodent. *J. Nutr.*, 131(11): 3082S-3086S.

Bogaards, J.J.P., Verhagen, H., Willems, M.I., van Poppel, G. and van Bladeren, P.J., 1994. Consumption of Brussels sprouts results in elevated α-class glutathione S-transferase levels in human blood plasma. *Carcinogenesis* 15: 5, 1073-1075.

Borgstrom, G., 1968. *Principles of Food Science. Vol. 2. Food Microbiology and Bio-chemistry*. The Macmillan Co. New York. 171-191.

Brewer, M.S, 1995. *Analyzing Food for Nutrition Labeling and Hazardous Contaminants* (Jeon, I.J. and Ikins, W.G., Eds.). Marcel Dekker, Inc. New York. 459-486.

Brown, V.K.H., 1981. *Testing for Toxicity* (Garrod, J.W., Ed.). Taylor and Francis Ltd. London. 21-43.

CAC, 1979. Guide to the safe use of food additives. *Codex Alimentarius*. 2nd series. FAO/WHO. Rome. 69 p.

CAC, 1983. Food Additives. *Codex Alimentarius*. Vol. XIV. 1st ed. FAO/WHO. Rome.1-29.

CAC, 1984. Contaminants. *Codex Alimentarius*. Vol. XVII. 1st ed. FAO/WHO. Rome. 33 p.

CAC, 1985. Report of the 18th session of the Codex Committee on Food Additives. FAO/WHO. Rome.

CAC, 1986. *Codex Alimentarius Commission Procedural Manual*. 5th ed. FAO/WHO. Rome. 40-41.

CAC, 1992. *Codex Alimentarius General Requirements*. Vol. 1. 2nd ed. FAO/WHO. Rome. 337 p.

CAC, 1997. Report of the 29th session of the Codex Committee on Food Additives and Contaminants. FAO/WHO. Rome. 57-180.

CAC, 1999a. Report of the 30th session of the Codex Committee on Food Additives and Contaminants. Codex Alimentarius Commission. Joint FAO/WHO Food Standards Programme. Alinorm 9912A. FAO/WHO. Rome. 469s.

CAC, 1999b. Report of the 31st session of the Codex Committee on Food Additives and Contaminants. Codex Alimentarius Commission. Joint FAO/WHO Food Standards Programme. Alinorm 9912A. FAO/WHO. Rome. 152s.

Carr, B.I., 1985. Chemical carcinogens and inhibitors of carcinogenesis in the human diet. *Cancer*, Vol. 55: 218-224.

Casarett, L.J., 1975. *Toxicology — The Basic Science of Poisons* (Casarett, L.J. and Doull, J., Eds.). Macmillan Publishing Co., Inc. New York. 3-10.

Chen, J., Montanari, A.M. and Widmer, W., 1997. Two new polymethoxylated flavones, a class of compounds with potential anticancer activity, isolated from cold pressed Dancy tangerine peel oil solids. *J. Agric. Food Chem.* 45: 364-368.

Chin, S.F., Liu, W., Storkson, J.M., Ha, Y.V. and Pariza, V.W., 1992. Dietary sources of conjugated dienoic isomers of linoleic acid, a newly recognized class of anticarcinogens, *J. Food Comp. and Anal.*, 5: 185-197.

Chu, F.S., 1995. *Analyzing Food for Nutrition Labeling and Hazardous Contaminants* (Jeon, I.J. and Ikins, W.G., Eds.). Marcel Dekker Inc. New York. 283-332.

Chuang, S.E., Kuo, M.L., Hsu, C.H., Chen, C.R., Lin, J.K., Lai, G.M., Hsieh, C.Y. and Cgheng, A.L., 2000. Curcumin-containing diet inhibits diethylnitrosamine-induced murine hepatocarcinogenesis. *Carcinogenesis*, 21:2, 331-335.

Concon, J.M., 1988$_a$. *Food Toxicology — Principles and Concepts. Part A*. Marcel Dekker, Inc. New York. 1-43; 155-239; 281-461.

Concon, J. M., 1988$_b$. *Food Toxicology-Contaminants and Additives. Part B*. Marcel Dekker, Inc. New York. 677-770; 1033-1325.

Considine, D.M. and Considine, G.D., (Eds.), 1982. *Foods and Food Production Encyclopedia*. Van Nostrand Reinhold Comp., Inc. New York. 45-47.

Coulson, J., 1980. *Developments in Food Colors*. (Walford, J., Ed.). Applied Science, London. 70; 214-218.

Derelanko, M.J., 1995. *CRC Handbook of Toxicology* (Derelanko, M.M. and Hollinger, M.A., Eds.). CRC. Boca Raton. 657. 6.

Doll, R., Peto, R., 1981. The causes of cancer. *J. Natl. Cancer Inst.*, 66, 1195-1308.

Dorant, E., Brandt, P.A., Goldbohm, R.A. and Sturmans. F., 1996. Consumption of onions and a reduced risk of stomach carcinoma. *Gastroenterology*, 110:12-20.

Durian, D.J. and Weitz, D.A., 1994. *Kirk-Othmer Encyclopedia of Chemical Technology* (Kroschwitz, J.C., Ed.). 4th ed. Wiley-Interscience. New York. 805-833.

EC, 1983. *Food-Science and Techniques. Reports of the Scientific Committee for Food*. 14th series. Office for Official Publications of the European Communities. Luxembourg. 41-46.

EC, 1988. Council Directive on the approximation of the laws of the member states relating to flavorings for use in foodstuffs and to source materials for their production. *Official J. Europ. Comm.*, No: 1, 184. 61-67.

EC, 1989. Council Directive on the approximation of the laws of the member states concerning food additives authorized for use in foodstuffs intended for human consumption, 89/107/EEC, *Official J. Europ. Comm.*, No: L 40. 27-33.

EC, 1990a. Proposal for a Council Regulation (EEC) laying down Community procedures for contaminants in food. Commission of the European Communities. Brussels. 9p.

EC, 1990b. Proposal for a council directive on sweeteners for use in foodstuffs. *Official J. Europ. Comm.* No. C 242. 4-14.

EC, 1992. Proposal for a council directive other than colors and sweeteners, 92/c 206/03, *Official J. Europ. Comm.*, No: C 206. 12-40.

EC, 1995. European Parliament and Council Directive, No: 95/2/EC, *Official J. Europ. Comm.*, No: L 61. 40 p.

Elias, P.S., 1971. *Sweetness and Sweeteners* (Birch, G.G., Green, L.F. and Coulson, C.B., Eds.). Applied Science Publishers. London. 139-159.

Fischetti, F. Jr., 1990. *CRC Handbook of Food Additives* (Furia,T.E. Ed.). Vol. 2. CRC,. Boston. 229-255.

Fontham, E.T.H., 1994. *Natural Antioxidants in Human Health and Disease* (Frei, B., Ed.). Academic, San Diego. 157-197.

Francis, F.J., 1985. *Food Chemistry* (Fennema, O.R., Ed.) 2nd ed. Marcel Dekker, 545-583.

Francis, F.J., 1992. *Natural Food Colorants* (Hendry, G.A.F. and Houghton, J.D., Eds.). Blackie and Son Ltd. London. 246-248.

Friedman, L.J. and Greenwald, C.G., 1994. Kirk-Othmer *Encyclopedia of Chemical Technology* (Kroschwitz, J.I., Ed.). 4th ed. Vol. 11. Wiley-Interscience. New York. 805-833.

Fürhacker, M., Scharf, S. and Weber, H., 2000. Bisphenol A: emissions from point sources. *Chemosphere*. Vol. 41, Issue 5. 751-756.

Garland, M., Stampfer, M.J., Willett, W.C. and Hunter, D.J., 1994. *Natural Antioxidants in Human Health and Disease* (Frei, B., Ed.). Academic, San Diego. 263-286.

Gerster, H., 1992. Anticarcinogenic effect of common carotenoids. *J. Vit. Nutr. Res.*, 63: 93-121.

Gomaa, E.A., Gray, J.I., Rabie, S., Lopez-bote, C. and Booren, A.M., 1993. Polycyclic aromatic hydrocarbons in smoked food products and commercial liquid smoke flavorings. *Food Additives and Contaminants*, Vol. 10, No. 5, 503-521.

Gormley, T.R., Downey, G. and O'Beirne, D., 1987. *Food Health and the Consumer*. Elsevier Applied Science. London. 26-63.

Harbison, R.D., 1980. *Casaret and Doull's Toxicology, The Basic Science of Poisons* (Doull, J., Klaassen, C.D. and Amdur, M.O., Eds.). 2nd ed. Macmillan, New York. 158-168.

Heath, H.B. and Reineccius, G., 1986. *Flavor Chemistry and Technology*. Van Nostrand Reinhold Comp. New York. 339-405.

Henry, B.S., 1992. *Natural Food Colorants*. (Hendry, G.A.F. and Houghton, J.D., Eds.) Blackie and Son Ltd. London. 39-78.

Hermus, R.J.J., Verhagen, H. and van Poppel, G., 1994. *Biomarkers in Nutritional Assessment. New Aspects of Nutritional Status* (Somogyi, J.C., Elmadfa, I. and Walter, P., Eds.). Basel. Bibl. Nutr. Dieta, Basel, Karger, No: 51, 116-125.

Hertog, M.G.L., Hollman, P.C.H. and Kaan, M.B., 1992. Content of potentially anticarcinogenic flavonoids of 28 vegetables and fruits commonly consumed in The Netherlands. *J. Agric. Food Chem.*, 40, 2379-2383.

Holt, P.R. and Miller, G.D., 1999. Calcium and prevention of chronic diseases. *J. Am. Coll. Nutr.* 18: Supplement 5, 379S-391S.

Homler, B.E., 1984. *Aspartame, Physiology and Biochemistry* (Stegink, L.D. and Filer, L.J., Eds.). Marcel Dekker. Inc. New York. 247-262.

Hutchings, J.B., 1977. *Sensory Properties of Foods* (Birch, G.G., Brennan, J.G. and Parker, K.J., Eds.). Applied Science Publishers Ltd. London. 45-57.

İçibal, N. and Altuğ, T., 1992. Degradation of aflatoxins in dried figs by sulphur dioxide alone and in combination with heat, ultraviolet energy and hydrogen peroxide. *Lebens. Wiss. U. Techn.*, 25, 294-296.

ICMSF, 1996. *Microorganisms in Foods 5. Characteristics of Microbial Pathogens*. Blackie Academic & Professional. London. 1st. ed. 20-35.

Inano, H., Onoda, M., Inafuku, N., Kubota, M., Kamada, Y., Osawa, T., Kobayashi, H. and Wakabayashi, K., 1999. Chemoprevention by curcumin during the promotion stage of tumorigenesis of mammary gland in rats irradiated with gamma-rays. *Carcinogenesis*, 20:6, 1011-1018.

Inano, H., Onoda, M., Inafuku, N., Kubota, M., Kamada, Y., Osawa, T., Kobayashi, H. and Wakabayashi, K., 2000. Potent preventive action of curcumin on radiation-induced initiation of mammary tumorigenesis in rats. *Carcinogenesis*, 21:10, 1835-1841.

Inglett, G.E., 1984. *Aspartame, Physiology and Biochemistry*. (Stegink, L.D. and Filer, L.J., Eds.). Marcel Dekker. Inc. New York. 11-25.

Jaffé, W.G., *Toxic Constituents of Plant Foodstuffs* (Liener, I.E., Ed.). 2nd ed. Academic San Francisco. 73-102.

Johnson, A.H. and Peterson, M.S., 1974. *Encyclopedia of Food Technology*. Avi. Connecticut. 7-11.

Johnson, E.A., 1990. *Foodborne Diseases* (Cliver, D.O., Ed.). Academic New York. 127-135.

Jones, J.M., 1992. *Food Safety*. Eagan Press. Minnesota. 31-52; 69-105; 141-169; 203-299; 331-417.

Kamrin, M.A., 1988. *Toxicology*. Lewis Publishers, Inc. Michigan. 1-78.

Karakaya, S. and Kavas, A., 1999. Adsorption of direct-acting and indirect acting mutagens by various dietary fibers. *Int. J. Food Sci. and Nutr.*, 50, 319-323.

Karakaya, S. and Kavas, A., 1999. Antimutagenic activities of some foods. *J. Sci. Food Agric.*, 79: 237-242.

Karapinar, M. and Gönül, S., 1999. Gida Mikrobiyolojisi (Ünlütürk, A. and Turantas, F., Eds.). Mengi Tan Basimevi. Izmir. 41-152.

Klaassen, C., 1980. *Casaret and Doull's Toxicology, The Basic Science of Poisons* (Doull, J., Klaassen, C.D. and Amdur, M.O., Eds.). 2nd ed. Macmillan, New York. 28-55.

Klaassen, C.D. and Doull, J., 1980. *Casaret and Doull's Toxicology, The Basic Science of Poisons* (Doull, J., Klaassen, C.D. and Amdur, M.O., Eds.). 2nd ed., Macmillan, New York. 11-15.

Kilgore, W. W. and Li, M., 1980. *Casaret and Doull's Toxicology, The Basic Science of Poisons* (Doull, J., Klaassen, C.D. and Amdur, M.O., Eds.). 2nd ed., Macmillan, New York. 593-607.

Kirk, R.S., 1980. *Food and Health: Science and Technology*. Applied Science. London. 287-291.

Knekt, P., 1994. *Natural Antioxidants in Human Health and Disease* (Frei, B., Ed.). Academic, San Diego. 199-238.

Krinsky, N.I., 1994. *Natural Antioxidants in Human Health and Disease* (Frei, B., Ed.). Academic, San Diego. 239-261.

Langseth, L., 1995. Oxidants, antioxidants and disease prevention. *ILSI Europe*. Belgium. 7-13.

Larsen, J.C. and Poulsen, E., 1987. *Toxicological Aspects of Food* (Miller, K., Ed.). Elsevier Science, London. 205-252.

Lau, B.H.S., Tadi, P.P. and Tosk, J.M., 1990. Allium sativum and cancer prevention. *Nutr. Res.*, 10: 937-948.

Lee, F.A., 1983. *Basic Food Chemistry*. 2nd ed. Avi, Connecticut. 237-260.

Lee, H., Jiaan, C.Y. and Tsai, S.J., 1992. Flavone inhibits mutagen formation during heating in glycine/creatine/glucose model system. *Food Chem.*, 45: 235-238.

Lessoff, M.H., 1998. Food allergy and other adverse reactions to food. ILSI Europe Concise Monograph Series. International Life Sciences Institute. Belgium. 22 p.

Lindsay, R.C., 1985. *Food Chemistry* (Fennema, O.R., Ed.). 2nd ed. Marcel Dekker, New York, 585-681.

Lipinski, G.W.R., 1988. *Low-Calorie Products* (Birch, G.G. and Lindley, M.G., Eds.). Elsevier Applied Science. New York. 101-112.

Lipkin, M., Newmark, H.L. and Miller, G.D., 1999. Calcium and prevention of chronic diseases. *J. Am. Coll. Nutr.* 18: Supplement 5, 392S-397S.

Lück, E. and Lipinski, G.W.R., 1988. *Ullmann's Encyclopedia of Industrial Chemistry* (Gerhartz, W., Ed.). Vol. A11. VCH. Verlagsgeselschaft mbH. Wienheim FRG. 561-581.

Luckey, T.D., 1972. *Handbook of Food Additives* (Furia, T. E., Ed.). Vol. I. 2nd ed. CRC, Florida. 1-26.

Machlin, L.J., 1991. *Handbook of Vitamins.* Marcel Dekker, New York 1. 43-133.

Malaspina, A., 1987. *Toxicological Aspects of Foods* (Miller, K., Ed.). Elsevier Applied Science, London. 17-57.

Malizio, C.J. and Johnson, E.A., 1991. Evaluation of the botulism hazard from vacuum packed enoki mushrooms (*Flammulina velutipse*). *J. Food Protect.*, 54:20-21.

Marie, S., 1991. *Food Additive User's Handbook* (Smith, J., Ed.). Blackie and Son, London. 46-74.

Martin, S.P. and Johnson, A.H., 1978. *Encyclopedia of Food Science.* Avi, Connecticut. 279-284; 348-350.

Marth, E.H., 1990. *Foodborne Diseases* (Cliver, D.O., Ed.). Academic, San Diego. 137-157.

Midorikawa, K., Murata M, Oikawa S., Hiraku, Y. and Kawanishi, S., 2001. Protective effect of phytic acid on oxidative DNA damage with reference to cancer chemo-prevention. *Biochem. Biophy. Res. Commun.*, 2; 288(3):552-557

Miller, E.C. and Miller, J.A., 1986. Carcinogens and mutagens that may occur in foods. *Cancer,* 58:1795-1803.

M. deMan, J. 1990. *Principles of Food Chemistry.* Avi, Connecticut. 414-455.

Mungia-Lopez, E.M. and Soto-Valdez, H., 2001. Effect of heat processing and storage time on migration of bisphenol A (BPA-diglycidyl ether [BADGE]) to acqueous food simulant from Mexican can coatings. *J. Agric. Food Chem.*, 49, 3666-3671.

Nabors, L. and Gelardy, R.C., 1991. *Handbook of Sweeteners* (Marie, S. and Piggott, J.R., Eds.). Blackie and Son, London. 104-115.

Neal, R.A., 1980. *Casaret and Doull's Toxicology, The Basic Science of Poisons* (Doull, J., Klaassen, C.D. and Amdur, M.O., Eds.). 2nd ed. Macmillan, New York. 56-69.

Noonan, J., 1990. *CRC Handbook of Food Additives* (Furia, T. E., Ed.). CRC, Vol. 1. Boston. 587-615.

Pals, I. and Jones, J.B., 1997. *The Handbook of Trace Elements.* St. Lucie, Boca Raton, Florida. 85-87; 93-95; 115-116; 120-122.

Pariza, 1996. *Present Knowledge in Nutrition* (Ziegler, E.E. and Filer, L.J., Eds.). Washington D.C., ILSI Press, 563-573.

Payner, P., 1991. *Food Additive User's Handbook* (Smith, J., Ed.). Blackie and Son, London. 90-113.

Perlman, F., 1980. *Toxic Constituents of Plant Foodstuffs* (Liener, I.E., Ed.). 2nd ed. Academic San Francisco. 295-327.

Peters, P.W., 1998. Developmental toxicology: adequacy of current methods. *Food Additives and Contaminants*, Vol. 15, Supplement, 55-62.

Pittet, A., 1998. Natural occurrence of mycotoxins in foods and feeds — an updated review. *Revue de Médécine Vétérinaire,* 149, 6, 479-492.

Potter, N.N., 1968. *Food Science.* Avi, Connecticut. 588-600.

Proudlove, R.K., 1989. *The Science and Technology of Foods*. Forbes Publications, London. 86-94.

Quattrucci, E., 1987. *Toxicological Aspects of Foods* (Miller, K., Ed.). Elsevier Applied Science, New York. 103-138.

Quilliam, M.A., 1995. Seafood toxins. General referee reports: *J. AOAC Int.*, Vol. 78, No: 1, 144-147.

Reddy, A.P., Harttig, U., Barth, M.C., Baird, W.M., Schimerlik, M., Hendricks, J.D. and Bailey, G.S., 1999. Inhibition of dibenzo(a,l)pyrene-induced multi-organ carcinogenesis by dietary chlorophyllin in rainbow trout. *Carcinogenesis*, 20:10, 1919-1926.

Reddy, C.S. and Hayes, A.W., 1989. *Principles and Methods of Toxicology* (Hayes, A.W., Ed.). Raven, 2nd ed. New York. 67-111.

Robinson, C.H., Lawler, M.R.,Chenoweth, W.L. and Garwick, A.E., 1986. *Normal and Therapeutic Nutrition*. Macmillan, New York. 76-77, 133, 496- 498.

Roe, F.J.C., 1981. *Testing for Toxicity* (Garrod, J.W., Ed.). Taylor and Francis, London. 29-43.

Roe, F.J.C., 1987. *Toxicological Aspects of Foods* (Miller, K., Ed.). Elsevier Applied Science. New York. 59-72.

Rompelberg, C.J.M., Bruijntjes-Rozier, G.C.D.M. and Verhagen, H., 1994. Effect of liver S9 prepared from eugenol-treated rats on the mutagenicity of benzo(a)pyrene and dimethylbenzanthracene. *Proc. Int. Conf. Euro Food Tox IV Bioactive Substances in Foods of Plant Origin* (Kozlowska, H., Fornal, J. and Zdunczyk, Z., Eds.). Centre for Agr. and Vet Sci. Poland, 2: 519-523.

Rompelberg, C.J.M., Verhagen, H. and van Bladeren, P.J.,1993. Effects of the naturally occurring alkylbenzenes eugenol and trans-anethole on drug-metabolizing enzymes in the rat liver. *Food Chem. Toxicol.*, 31:9, 637-645.

Saltmarsh, M. (Ed.), 2000. *Essential Guide to Food Additives*. LFRA Ltd., Leatherhead, Surrey. 1st ed. 1-59.

Schone, 1987. The Food Additives Policy in the European Community. Euro Coop. Brussels.1-14.

Schultze-Mosga, M.H., Dale, I.L., Gant, T.W., Chipman, J.K., Kerr, D.J. and Gescher, A., 1998. Regulation of c-fos transcription by chemopreventive isoflavonoids and lignans in MDA-MB-468 breast cancer cells. *Eur. J. Cancer*, Aug; 34 (9): 1425-1431.

Shao, Z., Jiong, W., Shen, Z., Barsky, S.H., Shao, Z.M., Wu, J. and Shen, Z.Z., 1998. Genistein inhibits both constitutive and EGF-stimulated invasion in ER-negative human breast carcinoma cell lines. *Anticancer-Research*, 18: 3A, 1435-1440.

Sharoni, Y. and Levy, J., 1996. *Natural Antioxidants and Food Quality in Atherosclerosis and Cancer Prevention* (Kumpulainen, J.T., Salonen, J.T., Eds.). The Royal Society of Chemistry. Cambridge, UK. 378-385.

Shills, M.E., 1986. *Modern Nutrition in Health and Disease*. (Shills, M.E. and Young, V.R., Eds.). Lea and Feibiger. Philadelphia. 1380-1387.

Sing, V.N. and Gaby, S.K., 1991. Premalignant lesions: role of antioxidant vitamins and beta-carotene in risk reduction and prevention of malignant transformation. *Am. J. Clin. Nutr.* Vol. 53, 386S-390S.

Snowstorm, L.B. and Little, A. D., 1990. *Handbook of Food Additives* (Furia, T.E., Ed.). Vol. 1. CRC, Boca Raton. 515-521.

Smith, J.S., 1995. *Analyzing Food for Nutrition Labeling and Hazardous Contaminants* (Jeon, I.J. and Ikins, W.G., Eds.). Marcel Dekker, New York. 333-361.

Solomon, H.C.M.C., Kautter, D.A., Lilly, T. and Rhodehamel, E., 1990. Outgrowth of *Clostridium botulinum* in shredded cabbage at room temperature under modified atmosphere. *J. Food Protection*, 53:831-833.

Stavric, B., 1994. Antimutagens and anticarcinogens in foods. *Food Chem. Toxicol.*, Vol. 32: No. 1, 79-90.

Suarez, S., Suerio, R.A. and Garrido, J., 2000. Genotoxicity of the coating lacquer on food cans, bisphenol A diglycidyl ether (BADGE), its hydrolisis products and a chlorohydrin of BADGE. In *Mutation Research/Genetic Toxicology and Environmental Mutagenesis*. Elsevier Science B.V., Vol. 470. Issue 2. 221-228.

Sugiyama, H., 1990. *Foodborne Diseases* (Cliver, D.O., Ed.). Academic New York. 107-125.

Sugiyama, H. and Yang, K. H., 1975. Growth potential of *Clostridium botulinum* in fresh mushrooms packaged in semipermeable plastic film. *Appl. Microbiol.*, 30: 964-969.

Sunshine, I. (Ed.), 1969. *CRC Handbook of Analytical Toxicology*. The Chemical Rubber Company. Ohio. 497-584.

Swaine, R. L. 1990. *CRC Handbook of Food Additives* (Furia, T.E., Ed.). CRC, Vol. 1. Boston. 457- 474.

Tartakow, I.J. and Vorperian, J.H., 1981. *Foodborne and Waterborne Diseases*. Avi, Westport, Connecticut, p. 213.

Taylor, S.L., 1990. *Foodborne Diseases* (Cliver, D.O., Ed.). Academic, San Diego. 17-43.

Taylor, S.L. and Schantz, E.J., 1990. *Foodborne Diseases* (Cliver, D.O., Ed.). Academic, San Diego. 67-84

Thiagarajan, D.G., Bennink, M.R., Bourquin, L.D. and Kavas, F.A., 1998. Prevention of precancerous colonic lesions in rats by soy flakes, soy flour, genistein, and calcium. *Am. J. Clin. Nutr.* 68 (supply): 1394S-1399S.

Thinly, W.G. and Libber, H.L., 1980. *Casaret and Doull's Toxicology, The Basic Science of Poisons* (Doull, J., Klaassen, C.D. and Amdur, M.O., Eds.). 2nd ed. Macmillan, New York. 139-153.

Thompson, L.U., Orcheson, L., Rickard, S., Jenab, M., Serranio, M., Seidl, M. and Cheung, F., 1996. *Natural Antioxidants and Food Quality in Atherosclerosis and Cancer Prevention* (Kumpulainen, J.T. and Salonen, J.T., Eds.). The Royal Society of Chemistry. Cambridge, UK. 356-364.

Tolonen, M., 1990. *Vitamins and Minerals in Health and Nutrition*. Ellis Horwood, West Sussex, England. 88-98.

Toth, L., Potthast, K., 1984. *Advances in Food Research*. Vol. 29. Academic, New York.134-149.

Ueno, Y., 1987. *Toxicological Aspects of Food* (Miller, K., Ed.). Elsevier Applied Science, New York. 139-204.

USDA, 2002. Guidance for Industray — Preparation of Food contact Notifications and Food Additive Petitions for Food Contact Substances: Chemistry Recommendations. Final Guidance April 2002. USFDA Center for Food Safety and Applied Nutrition. Office of Food Additive Safety. Web site: www. cfsan.fda.gov/dms/opa2pmnc.html.

USDA, 2002. Guidance for Industry

Verhagen, H., de Vogel, N., Verhoef, A., Schouten, T. and Hagenaars, A.J.M., 1994. Modifying effects of garlic constituents on cyclophosphomide and mitomycin C-induced (Geno) toxicity in mouse bone marrow cells *in vivo*. *Proc. Int. Conf. Euro Food Tox IV. Bioactive Substances in Food of Plant Origin*, (Kozlowska, H., Fornal, J. and Zdunczyk, Z., Eds.). Centre for Agr. and Vet Sci., Poland. 2: 524-529.

Verhagen, H. and Feron, V.J., 1994. Cancer prevention by natural food constituents — the lessons of toxicology transposed to antigenotoxicity and anticarcinogenecity. *Proc. Int. Conf. Euro Food Tox IV Bioactive Substances in Food of Plant Origin* (Kozlowska, H., Fornal, J. and Zduńczyk, Z., Eds.). Centre for Agr. and Vet Sci., Poland, 2, pp: 463-478.

Verhagen, H., Poppel, M.I., Bogaards, J.J.J.P., Rompelberg, C.J.M. and Bladeren, P.J., 1993. Cancer prevention by natural food constituents. *Int. Food Ingredients*, No: 1/2, 22-29.

Verhagen, H., Rompelberg, C.J.M., Strube, M. van Poppel, G. and van Bladeren, P.J., 1997. Cancer prevention by dietary constituents in toxicological perspective. *J. Environ. Pathol., Toxicol. and Oncol.* 16 (4): 343-360.

Vettorazzi, G., 1987. *Toxicological Aspects of Foods* (Miller, K., Ed.). Elsevier Applied Science. New York. 1-16.

Walker, R., 1998. Toxicity testing and derivation of ADI. *Food Additives and Contaminants*, Vol. 15, Supplement, 11-16.

Walker, R., 1999. Development of toxicological test methods: possibilities and limits. ILSI Europe Seminar on Food Safety and Nutrition Policy: developments in safety assessment and nutrition science. *ILSI Europe*. Ankara. 56-65.

Wattenberg, L.W., 1985. Chemoprevention of cancer. *Cancer Research*, 45, 1-8.

Weisburger, J.H. and Williams, G.M., 1980. *Casaret and Doull's Toxicology, The Basic Science of Poisons* (Doull, J., Klaassen, C.D. and Amdur, M.O., Eds.). 2nd ed. Macmillan, New York. 84-87.

Whanger, P.D., 1982. *Nutritional Toxicology* (Hatchcock, J.N., Ed.). Vol. 1. Academic Press. New York. 163-208.

WHO, 1978. Principles and methods for evaluating the toxicity of chemicals. Part 1. Environmental Health Criteria 6. WHO. U.K. 1-27; 95-115.

WHO, 1987. Principles for the safety assessment of food additives and contaminants in food. Environmental Health Criteria. No: 70. WHO. Geneva. 174 p.

Wodicka, V.O., 1990. *CRC Handbook of Food Additives* (Furia, T.E., Ed.). 2nd ed. Vol. I. CRC, Florida. 1-12.

Yen, G.C. and Chen, H.Y., 1994. Antimutagenic effect of various tea extracts. *J. Food Prot.* 7: 1, 54-58.

York, G.K. and Gruenwedel, S.H.O., 1990. *Chemicals in the Human Food Chain* (Winter, C., Seiber, J.N. and Nuckton, C.F., Eds.). Van Nostrand Reinhold. New York. 87-128.

Yoshitani, S.I., Tanaka, T., Kohno, H. and Takashima, S., 2001. Chemoprevention of azoxymethane-induced rat colon carcinogenesis by dietary capsaicin and rotenone. *Int. J. Oncol.*, 19 (5):929-939.

Xing, M., 1982. Garlic and gastric cancer — the effect of garlic on nitrite and nitrate in gastric juice. *Acta Nutrimenta Sinica* 4: 1, 53-58.

Zheng, G., Kenney, P.M. and Lam, L.K.T., 1992. Myricticin: A potential cancer chemopreventive agent from parsley leaf oil. *J. Agric. Food Chem.*, 40: 107-110.

INDEX

Gallic acid esters (dodecyl, octyly,
        and propyl gallates), 91
Gambierdiscus toxicus, 48
Garlic, 91, 116, 117, 124
Gastric cancer, 64, 116, *see also*
        Stomach cancer
Gastroenteritis, 58
Gelatin, 77, 89, 92, 97, 98, 101
Gelling agents, 83, 98
GEMS/Food, 70
Gene-locus mutations *see* Point
        mutations
Genetic toxicology, 5
Genistein, 119, 120
Genotoxic carcinogens, 25, 27
Genotoxicity, 35, 38, 109
German measles, 26, *see also* Rubella
Glazing agents, 84, 98
Glucomannan, 117
Gluconic acid, 89
δ-Gluconolactone, 101
Glucose oxidase, 91
Glucosinolates, 47, 110, 123
Glucose-6-phosphate dehydrogenase,
        45
Glutamates 97, 98
Glutamine, 20
Glutathione, 112
Glutathione-S-transferase, 107, 110,
        112, 123
Glycine, 20, 118
GMP, 87, 88
Goiter
Goitrogens, 47, *see also*
        Glucosinolates
Gonyaulax
    catenella, 50
    tamarensis, 50
Gonyautoxin
    II, 50
    III, 50
Good manufacturing practice, *see*
        GMP
Gossypol, 10, 28, 47, 106
GRAS substances, 82, 102

Green S, 93
Guanylic acid, potassium and
        calcium salts, 98
Guar gum, 101, 104
Gum arabic, 101
Gyromitra esculentaI, 46, *see also*
        Hydrazines
Gyromitrin, 46

H

HACCP (hazard analysis critical
        control points), 54
Halogenated hydrocarbon pesticides,
        69
Hazard characterization, 37, 38
Heavy metals, 28, 79, 114
Hemaglutinins, 47
Hematuria, 45
Hemolytic anemia, 45
Hemicelluloses, 117
Hemoglobin, 18, 44
Hepatic cytochrome P-450
        monooxygenase system,
        19, 20
Hepatitis B virus, 20
Heptachlor, 9, 69, 70
Herbs, 28, 58, 95
Herbicides, 10, 69, 70–76
Heterocyclic amines, 24
Histamine, 49, 52
Histidine, 52
Hormones, 1, 27–29, 59
HT-2 Toxin, 64
Human Studies, 35, 121, 125, *see*
        also Epidemiological
        studies and Biomarker
        research
Humectants, 82, 84, 99
Hydrazines, 28, 46, 106
Hydrocolloids, 101, 103
Hydrogen
    peroxide, 24, 61
    sulfide, 10
Hydrogenated glucose syrups, 102